Studies in Applied Philosophy, Epistemology and Rational Ethics

Volume 30

About this Series

Studies in Applied Philosophy, Epistemology and Rational Ethics (SAPERE) publishes new developments and advances in all the fields of philosophy, epistemology, and ethics, bringing them together with a cluster of scientific disciplines and technological outcomes: from computer science to life sciences, from economics, law, and education to engineering, logic, and mathematics, from medicine to physics, human sciences, and politics. It aims at covering all the challenging philosophical and ethical themes of contemporary society, making them appropriately applicable to contemporary theoretical, methodological, and practical problems, impasses, controversies, and conflicts. The series includes monographs, lecture notes, selected contributions from specialized conferences and workshops as well as selected Ph.D. theses.

Advisory Board

More information about this series at http://www.springer.com/series/10087

Emily Rolfe Grosholz

Starry Reckoning: Reference and Analysis in Mathematics and Cosmology

Springer

Emily Rolfe Grosholz
Department of Philosophy
The Pennsylvania State University
University Park, PA
USA

ISSN 2192-6255 ISSN 2192-6263 (electronic)
Studies in Applied Philosophy, Epistemology and Rational Ethics
ISBN 978-3-319-46689-7 ISBN 978-3-319-46690-3 (eBook)
DOI 10.1007/978-3-319-46690-3

Library of Congress Control Number: 2016953651

Printed on acid-free paper

This Springer imprint is published by Springer Nature
The registered company is Springer International Publishing AG
The registered company address is: Gewerbestrasse 11, 6330 Cham, Switzerland

Preface

This book stems from my engagement with philosophers of mathematics at the University of Rome, the University of Paris, the University of London and King's College, London, and members of the newly formed Association for the Philosophy of Mathematical Practice, as well as mathematics seminars at the Pennsylvania State University, in particular those of the number theorist Wen-Ch'ing (Winnie) Li. It also stems from my membership in the Center for Fundamental Theory in the Institute for Gravitation and the Cosmos at Penn State. Recently I went over there to help celebrate the fact that for the past 15 years, the IGC has been one of the main incubators of the spectacular discovery of the gravitational waves that Einstein predicated a century ago. My membership as the resident Humanist began seven years ago, and resulted in a workshop on cosmology and time where philosophers and scientists engaged in fruitful conversation now collected in a special issue of *Studies in the History and Philosophy of Modern Physics*, Vol. 52, Part A (2015). One of the aspects of research at the IGC that struck me right away was the substantive collaboration between theoretical physicists and highly empirical astronomers, despite the striking disparity in the kinds of texts they generated. Watching their interchanges helped me to develop further ideas about mathematical and scientific research, which I began to formulate during a sabbatical leave supported by a fellowship from the City of Paris and of course also Penn State. So I am indebted to colleagues far away and close to home.

The workshop on cosmology and time took place with support from then Head of the Philosophy Department Shannon Sullivan and Dean Susan Welch, warm support from Abhay Ashtekar and Murat Gunaydin at the IGC and Professor of Physics Emeritus Gordon Fleming, as well as John Norton, Director of the Center for History and Philosophy of Science at the University of Pittsburgh. We listened to exchanges between Bryan Roberts and Abhay Ashtekar, William Nelson and Sarah Shandera, Thomas Pashby and Gordon Fleming, David Sloan and Kurt Gibble, Elie During and myself, and Alexis de Saint-Ours and John Norton; later additions included responses or essays by Jeremy Butterfield, Julian Barbour, Klaus Mainzer, and Lee Smolin, to complement the 'overview' essays by Abhay Ashtekar and John Norton that conclude the special issue of *SHPMP*.

The relation of the theoretical physicists to the astronomers in cosmological research seems to me analogous to the relation of the logicians to the number theorists, geometers and topologists I have encountered over the years at Penn State and elsewhere. Philosophy of mathematics for many decades has been dominated by logicians who (I think) misunderstand the role of mathematical logic. It is not an over-discourse that should supplant the others, but one of many, which can be integrated with other mathematical discourses in a variety of fruitful ways. Thus I am happy to note that philosophy of mathematics in the early 21st century is undergoing a long-awaited and long-overdue sea change. Philosophers of mathematics are turning to a serious study of the *history* of mathematics, logic, and philosophy, as well as interviewing mathematicians and educators, and reading textbooks, to look in detail at mathematical research and instruction. They are trying to give an account of how mathematical knowledge grows and how problems are solved, with the dawning insight that serious problem-solving is by nature ampliative. In addition to the founding of the Association for the Philosophy of Mathematical Practice in 2009, I would point to a series of publications initiated by Carlo Cellucci, and Brendon Larvor's Mathematical Cultures project (2011–2014), as well as Mic Detlefsen's Ideals of Proof project (2008–2011) and Karine Chemla's Mathematical Sciences in the Ancient World project (2011–2014), both of the latter conjured up in the wonderful research cauldron SPHERE (CNRS UMR 7219) at the University of Paris Denis Diderot—Paris 7. To this we can now add Springer's SAPERE series under the direction of Lorenzo Magnani.

In fact, this change had been in preparation for decades. Imre Lakatos' celebrated book *Proofs and Refutations: The Logic of Mathematical Discovery* was published in 1976, and inspired many young philosophers to think about mathematics in a new way. Donald Gillies wrote his dissertation with Lakatos, and has been active in the philosophy of mathematics since the late 1970s. He published my essay "Two Episodes in the Unification of Logic and Topology" in the *British Journal for Philosophy of Science* in 1985, and that same year introduced me to Carlo Cellucci when I was in Rome, who would go on to write three important books on logic and mathematics in Italian, in 1998, 2002, and 2007. (His *Rethinking Logic: Logic in Relation to Mathematics, Evolution, and Method* was published in English in 2013, followed shortly by Danielle Macbeth's similarly motivated but dialectically opposed *Realizing Reason: A Narrative of Truth and Knowing* in 2014.) In 1995, Donald Gillies edited an important collection of essays, *Revolutions in Mathematics*, thoroughly Lakatosian in spirit, with contributions by *inter alia* Paolo Mancosu, Herbert Breger, Caroline Dunmore, Jeremy Gray, Michael Crowe, Herbert Mehrtens, Joe Dauben, and myself. In 1994 and 1995, I organized two conferences at Penn State (with support from a Humboldt Foundation Transatlantic Cooperation grant written with Herbert Breger, then director of the Leibniz Archives), where philosophers and historians of mathematics interacted and wrote pairs of essays together, so that philosophical claims might be supported, countered or complicated by case studies from the history of mathematics, which was precisely what Lakatos had encouraged. This resulted in *The Growth of Mathematical Knowledge* in 1999, and included essays by Herbert

Breger and myself, Carlo Cellucci, Donald Gillies, Penelope Maddy, Paolo Mancosu, François De Gandt, Jaakko Hintikka, Madeline Muntersbjorn, Carl Posy, Mike Resnik, Hourya Sinaceur, Klaus Mainzer, and others.

In 2005, Carlo Cellucci edited, with Donald Gillies, *Mathematical Reasoning and Heuristics*; in 2006, with Paolo Pecere, *Demonstrative and Non-demonstrative Reasoning in Mathematics and Natural Science*; and in 2011, with myself and Emiliano Ippoliti, *Logic and Knowledge*, as well as a special issue of the Italian journal *Paradigmi*. The *Festschrift* for him edited in 2014 by Emiliano Ippoliti and Cesare Cozzo continues this series. In the Francophone world, the tradition of philosophy of mathematics informed by history of mathematics remained unbroken throughout the 20th century. Jules Vuillemin, for example, who was in a sense the Quine of France, did not share Quine's disdain for history; rather, he was an expert on the history of mathematics, logic and philosophy, as well as the latest texts in analytic philosophy. His book *La Philosophie de l'algèbre* (1962/1993) has been very influential, and so likewise his magisterial *Nécessité ou contingence: l'aporie de Diodore et les systèmes philosophiques*, published in 1984. He belonged to the tradition of Henri Poincaré, the great *fin de siècle* mathematician-physicist-philosopher; indeed, Vuillemin's papers are housed in the Archives Henri Poincaré at the University of Lorraine, under the auspices of Gerhard Heinzmann, who published *L'Intuition épistémique* in 2013. Also included in this lineage are Albert Lautman, Jean Cavaillès, and Léon Brunschvicg (and the German Ernst Cassirer), all of whom met their fate in 1944–1945. (Emmy Noether worked with Cavaillès on an edition of the Cantor-Dedekind correspondence.) Vuillemin was my philosophical mentor, and the teacher of Claude Imbert (whose *Pour une histoire de la logique* appeared in 1999) and Hourya Sinaceur (whose book on the theory of fields, *Corps et Modèles*, appeared in 1991). I met the Newton scholar François De Gandt in 1981 through Rémi Brague, when I attended a study group in Paris on Book I of the *Principia*; he introduced me to Hourya Sinaceur and Karine Chemla (she translated, with Guo Shuchun, the Chinese mathematical classic *Les neuf chapitres*), and Amy Dahan and Jeanne Pfeiffer (who together edited *Une histoire des mathématiques: routes et dédales*, (1986), which was published in English translation in 2009).

Almost all the French and Italian philosophers just mentioned were invited to the United States or Canada, and spent some months or a year or so there during the last two decades of the 20th century, interacting with many important philosophers and historians of mathematics. They in turn introduced me to a generation of younger scholars: Marc Parmentier, David Rabouin, Jean-Jacques Szczeciniarz, Christine Proust, Andy Arana, Dirk Schlimm, Agathe Keller, Renaud Chorlay, Valeria Giardino, Marco Panza, Ivahn Smadja, Norma Goethe, Amirouche Moktefi, Justin Smith, Valérie Debuiche, Koen Vermeir, Pierre Cassou-Noguès, Jessica Carter, Giulia Miotti, Roy Wagner, and many others associated with the University of Rome 'La Sapienza,' SPHERE and elsewhere. This vigorous, unbroken European tradition has played a big role in the development of a richer approach to philosophy of mathematics, whose proponents moreover now interact fruitfully with philosophers who continue to be more interested in logic and formalism, as recent conferences, lecture series and events attest, organized by Mic Detlefsen, Brendan Larvor, Lorenzo Magnani, José Ferreirós, Jamie Tappenden, Michel Serfati, Marco Panza, Penelope Maddy, Dale Jacquette, Marcus Giaquinto, Paolo Mancosu and others.

To this list should be added philosophically inclined mathematicians whose "popular" expositions of various issues in contemporary mathematics and the history of mathematics raise lively issues about mathematical reason, and add important dimensions to the current conversation: Joseph Mazur, Robyn Arianrhod, Reuben Hersh, William Byers, Edward Frenkel, Donal O'Shea, Marjorie Senechal and Chandler Davis. As noted above, I am indebted to the number theorist Winnie Li for allowing me to sit in on her graduate seminars over a period of many years, and to graduate students Travis Morrison, Sonny Arora, Ryan Flynn, William Chen, Ayla Gafni, and Haining Wang for sharing their research with me. And reiterated thanks, in the final preparation of the manuscript of the book, to Gordon Fleming, Winnie Li, Jeremy Gray, and Christelle Vincent.

I think it is important to recognize the sea change in philosophy of mathematics (following upon a change in philosophy of science that began with Kuhn), which has taken place over the last forty years, and to bring the scattered philosophers, mathematicians and historians responsible for it into more coherent, though of course still dialectical, conversation. This book is my contribution to the current conversation, which so often focuses on the nature of ampliative reasoning in mathematics and the mathematical sciences, in this case those where we count moments and stars as well as units. Hence the title of this book.

University Park, USA Emily Rolfe Grosholz

Contents

Introduction

My central concern in this book is the growth of knowledge: what happens when reasoning not only orders what we know, but adds to what we know? Deductive logic has traditionally offered important answers to the question of how reasoning orders what we know, and axiomatic systems like Euclid's *Elements* or Newton's *Principia* have served as exemplars for the way in which scattered results can be unified and brought into useful relation with each other. Inductive logic and abductive logic, explored in the writings of Mill and Peirce, have been used to address the question of how reasoning can add to our knowledge. However, since they were both logicians, there is a limit on their explorations of the question. Because of the nature of logic, a logician typically assumes that the discourse which expresses the kind of reasoning explored is homogeneous. If signs shift their meanings in the midst of a deduction, for example, we have an instance of the fallacy of equivocation, and the reasoning is invalid.

However, I claim that some of the most important ampliative reasoning—reasoning that extends knowledge—in mathematics and the sciences takes place when heterogeneous discourses are brought into novel, and rational, relation. The study of rationality is not limited to the discipline of logic, though of course it must include logic as one important aspect. However, my way of proceeding is first and foremost historical; the way to see and understand these kinds of rational processes is to track them through history, through the notes of mathematicians and physicists, their published papers, and the presentation of new developments in textbooks. Those who bring heterogeneous discourses together can proceed in a variety of ways. They can proceed by juxtaposition: thus in Descartes' *Geometry* we see geometrical diagrams side by side on the page with the old-fashioned idiom of proportions, and the new idiom of polynomial equations. This juxtaposition can be forced into superposition: thus in Newton's proof of Proposition XI in Book I of the *Principia*, we must understand the same line segment QR as a representation of force (the novel meaning given to it by Newton's resolution of the 2-body problem in terms of gravitational attraction) and as the first segment of the geometrical line articulated into units that illustrates naturally accelerated motion in Theorem II,

Proposition II, of the Third Day of Galileo's *Discorsi*, and exhibits the fact that the distance traversed is proportional to the square of the time elapsed; we must also read *QR* as both finite and infinitesimal for the argument to go through. And, finally, superposition can be forced into unification: the modification of the theory of proportions that Newton profits from is the treatment of all magnitudes (formerly and in other contexts regarded as heterogeneous and carefully segregated) as numbers, all ratios as fractions (which then also join the ranks of number) and all proportions as equations between numbers. In none of these cases does the heterogeneity disappear, but it is exploited differently. The virtue of heterogeneous discourse is that investigators can bring different kinds of information, conceptualizations, methods and formal idioms to bear *together* on problem-solving situations. Rather than trying to banish the heterogeneity, as logicians strive to do in both their mathematics and their philosophizing, I urge that we look philosophically, and historically, at the uses that mathematicians and scientists make of the heterogeneity of things and of discourses, and the strategies they use to bring them together.

In this book, I explore a certain range of this kind of reasoning, without claiming that this exhausts the kind of ampliative reasoning that interests me. Far from it: I'd be happy to inspire other philosophers to look into other patterns of reasoning between and among heterogeneous discourses, which with the advent of computer technology are multiplying, and diverging, quickly. To do this, we need to interact directly with mathematicians and scientists doing research at the various frontiers of mathematics and physics, and to work harder at mastering the history of those domains. In this book, my focus is on the following patterns of reasoning. I argue that productive mathematical and scientific research takes place when discourse whose main intent is to establish and clarify reference is yoked with discourse whose main intent is analysis. I borrow the term analysis here from Leibniz: it means both the search for conditions of solvability of problems, and the search for conditions of intelligibility of things. In mathematics, one instructive example is the disparity between the investigations of figures in geometry, and the analytic elaborations of algebra, the infinitesimal calculus and the theory of differential equations. Another is the disparity between the investigation of numbers in number theory, and the analytical elaborations of mathematical logic. Two more historical examples arise in cosmology: the disparity between Newtonian Mechanics and Thermodynamics in the 19th century and between General Relativity and Quantum Mechanics in the 20th century. And in the new century, we can track the disparity between the empirical discourse of astronomers and the abstract theories of modern cosmologists who pursue, for example, Quantum Gravity.

I don't see much difference between these cases; another way of putting the point is that I am not interested in entering the long-standing and vexed debate about the ontological status of mathematical things. A continuation of that debate will not show up in the pages of this book; there are no reflections on the 'reality' of numbers and figures, as opposed to that of moments of time or stars. Mathematicians refer to canonical mathematical things, and strive to expand their taxonomies and to make them more precise; so do astronomers with respect to astronomical things. In both cases, the discourses that allow them to refer well, and the discourses that allow them to analyze well, typically look very different. What interests me are the strategies employed to bring these discourses into working relationship; what I am trying to do is explain how and why these strategies work well, and extend our knowledge.

Here is a brief account of the plan of the book. In Chap. 1, I elaborate a bit on the opposition between referential discourse and analytic discourse, and then devote some pages to a more detailed account of Leibniz's notion of analysis, clarifying it in relation to his debates with Locke in the *Nouveaux Essais*, and then examining its continuation in the work of Ernst Cassirer. Cassirer, I argue, tends to read superposition too strongly as unification, as if the referential discourse disappeared entirely into the analytic discourse; to contest this reading, I tell a historical narrative about the investigation of the circle. The circle, like the cosmos, always has further surprises to reveal; its determinate oneness is never exhausted by an analytic discourse.

Having explained the issue that I want to address, in Chap. 2 I defend my method. My method is based on the study of history rather than on logic; logic plays a role but that role is subordinate or subsidiary. Philosophers are now used to the idea that history is central to philosophy of science, but its pertinence to the philosophy of mathematics seems to need renewed defense. Thus I rehearse the helpful argument of the philosopher of history, W. B. Gallie, who shows that there can be no Ideal Chronicle, which in turn helps me to contest Philip Kitcher's early and influential position in *The Nature of Mathematical Knowledge*, where he invokes the history of mathematics. But there are two assumptions that render Kitcher's account ahistorical. One is that mathematical knowledge has its origins in physical processes that cause fundamental beliefs in us (and these processes, while temporal, are not historical); this is a commitment to naturalized epistemology that he shares with Penelope Maddy. The other is that mathematics as it develops should ultimately be 'stabilized' as a unified, universal, axiomatized system where all problems are solved and have their place as theorems. This account leaves history behind, in both its empiricist origins and its logicist completion. Thus I turn to the French philosopher Cavaillès whose sense of history was more refined, make some remarks about reference, and take up briefly Wiles' strategy in his celebrated proof of Fermat's Last Theorem, in order to carry out a thought experiment that shows that proofs are embedded in history, like all human actions.

Chapter 3 further explores my methodological assumptions, and shows why in investigating ampliative reasoning, I put so much emphasis on analysis and its interplay with reference. I go back over my own engagement with recent work by Karine Chemla, Carlo Cellucci and Danielle Macbeth to defend my approach and choice of case studies in the chapters that follow, as well as work by two mathematics educators, Kieran and David Egan, in order to revisit my arguments about the circle, and to prepare the ground for my case studies in the next two chapters, where number theory is developed in relation to geometry, algebra, complex analysis, and topology. In Chap. 4, I review the Nagelian model of theory reduction, with its projection of a collapse of two discourses into one, and then Jeremy

Butterfield's argument that in physics definitional extension is both too strong and two weak to conform to Nagel's strict criteria, and so blocks the collapse. I then argue that the same holds true for attempts at theory reduction in mathematics, by tracking the expositions of alleged reductions in a logic textbook by H.B. Enderton. Finally, I trace the proof of a few of Fermat's conjectures by means of the alliance of number theory with complex analysis, and show that the truly ampliative and explanatory proofs open up the study of the rational numbers to the study of algebraic number fields, an extension of number theory that is once too strong and too weak to look like Nagelian theory reduction, which is precisely why it turns out to be so fruitful.

In Chap. 5, I look in some detail at Enderton's exposition of number theory in his logic textbook, and at Gödel's incompleteness theorems. Once again, I show that definitional extension is both too strong and too weak; the disparity and the resultant ambiguity in both the textbook and the proofs testify to the disparity between the referential discourse of arithmetic and number theory and the analytic discourse of logic. Then I turn to the preliminary stage of Andrew Wiles' proof of Fermat's Last Theorem. Fermat's Last Theorem follows from a two-way correspondence between elliptic curves and modular forms, because that correspondence rules out counterexamples to it. The Taniyama-Shimura conjecture, that every (semi-stable) elliptic curve over **Q** is modular, is what Wiles proved; the converse claim, that every modular form corresponds to a certain elliptic curve, had already been proved by Eichler and Shimura: that is the result I discuss in some detail, because it is a bit simpler to explain, and yet can exemplify and exhibit my argument about the whole proof. The proof itself combines disparate discourses, in a spectacularly ampliative and inspiring way. Then I go on to discuss attempts by the logicians McLarty, Friedman, and Macintyre to rewrite the proof. I try to show that the aims of logicians are different from the aims of number theorists: the 'logically extravagent' general methods and theorems of the proof seem to be needed to organize the practice of the number theorists, while the 'messy and piecemeal' methods of the logicians reveal aspects of the proof (its reducible logical complexity) central to the research practice of logicians (in this case, model theorists and category theorists). All the same, this further disparity is helpful for the development of mathematics; it can contribute to the growth of knowledge as long as both sides tolerate each other and remain open to novel kinds of interaction, with neither claiming to have the Ultimate Discourse.

This is an insight that applies just as well to the sciences. I have explained to a number of cosmologists my reasons for thinking that there is no Grand Unified Theory, and why indeed that might be a misleading ideal. (I was met with skepticism.) In the last three chapters of the book, I look at the history of cosmology, from 17th c. debates to contemporary developments, with special emphasis on investigations into the nature of time. In Chap. 6, I review the innovative mathematical representations of time in the work of Galileo, Descartes and Newton, and then turn to the debate over whether time is absolute (to be defined analytically) or relative (to be defined referentially) between Newton and his mouthpiece Clarke, and Leibniz. In this process, I re-examine in more detail Leibniz's treatment of time.

This is really an exercise in and rehearsal of methodology, as were my initial chapters, also inspired by Leibniz. How shall we think about the ways in which the two kinds of discourse, empirical compilation and theoretical analysis, may be combined in science? Leibniz calls on metaphysics, in particular the Principle of Sufficient Reason, to regulate a science that must be both empirical and rationalist: the correlation of precise empirical description with the abstract conception of science *more geometrico* is guaranteed by the thoroughgoing intelligibility and perfection of the created world. Leibniz encourages us to work out our sciences through successive stages, moving back and forth between concrete taxonomy and abstract systematization, and this model of scientific inquiry accords very well, as I try to show, with his own investigations into mechanics and planetary motion, and so too his mathematical-metaphysical account of time, which was subtler and more multivalent than that of Newton.

I also review the influence of Leibniz on the contemporary cosmologists Penrose and Smolin, in the transition to Chap. 7, which examines debates over the nature of time in recent centuries. I look at the revision of the concept of time occasioned by 19th century thermodynamics, and then Boltzmann's attempt to reconcile it with Newtonian mechanics: is the arrow of time (so referentially compelling) real, or can it be explained away by an analytic discourse? During the 20th century, in a sense classical General Relativity Theory continued the Newtonian tradition of an analytic, geometrical theory of time, and Quantum Mechanics continued the Leibnizian tradition of a referential theory of time elicited from the dynamical object (molecular, atomic, and subatomic particles); and the dialectic, modified, continues into the current century. My view is that the heterogeneity of the discourses and their complementarity are useful for the advance of science; and that the interesting philosophical question is how the two approaches interact.

The last chapter, Chap. 8, looks into the use of reference and analysis in the study of astronomical systems from Newton to the present day. I begin by noting that whereas Bas van Fraassen pays more attention to theoretical models that do the work of analysis, Nancy Cartwright and Margaret Morrisson pay more attention to models where the relevant relation is not satisfaction (as between a meta-language and an object-language), but representation (as between a discursive entity and a thing that exists independent of discourse), like the iconic images that represent molecules. I then track the use of both kinds of models in Newton's *Principia*, Books I and III, and find strategies of juxtaposition, superposition, and unification nuanced by both. Then I argue that in the era after Newton, problems of analysis were addressed by Euler, Lagrange, Laplace and Hamilton, while problems of reference came to the fore in the empirical work of Herschel and Rosse. In the same spirit, I look at the debates between Hubble and Zwicky, ending with a reflection on the work of Vera Rubin, whose empirical investigations led to a most esoteric theoretical conjecture, with which cosmologists still struggle to come to terms.

The advantage of analytic discourse is that it is great for organization, indexing and generalization; however, it also tends to unfocus the specificity of things, to make them ghostly. The advantage of referential discourse is that it does better justice to the rich, specific variety of things, but it often loses its way in the forest of research because of all those trees. This complementarity holds just as much in mathematics as it does in the sciences. What strategies do we use to bring these discourses into rational, productive relation? To what extent must we violate deductive logic, tolerate a multiplicity of theories, and exploit well-structured ambiguity to do this, and what does this mean for our conception of rationality? I hope that my book will advance our understanding of reason.

Chapter 1
Reference and Analysis

Productive scientific and mathematical discourse must carry out two distinct tasks in tandem: an analysis or search for conditions of intelligibility (things) and solvability (problems); and a strategy for achieving successful reference, the clear and public indication of what we are talking about. More abstract discourses that promote analysis, and more concrete discourses that enable reference are typically not the same, and the resultant composite text characteristic of successful mathematical and scientific research will thus be heterogeneous and multivalent, a fact that has been missed by philosophers who begin from the point of view of logic, where rationality is often equated with strict discursive homogeneity, and from empiricism in its recent version of 'naturalized epistemology', focusing on the formation of beliefs from sensory stimulation. Thus I investigate how and why problem-solving may be ampliative in mathematics and physics, increasing the content of knowledge and crossing boundaries. And I look at the investigation of numbers, time and stars, three kinds of things that cannot be directly perceived (with the exception of the sun, and about ten thousand stars visible to the human eye). My arguments are supported by case studies from the early modern period, the 19th century and the more recent past, since I think that historical resemblances and differences make my philosophical case stronger.

Concepts of rationality and method have been central to philosophy of mathematics, philosophy of science, and epistemology, since Descartes wrote his *Discourse on Method*, *Geometry* and *Meditations* at the beginning of the 17th century. During the early 20th century, however, philosophers have all too often equated rationality with logic, and method with the organization of knowledge into axiomatic systems expressed in a formalized language. Since valid deductive argument forms guarantee the transmission of truth from premises to conclusion, deductive logic is important as a guide to efficacious mathematical and scientific reasoning; however, it does not exhaust them. But valid deductive reasoning requires discursive homogeneity; apples cannot be deduced from oranges. Thus an unswerving focus on logic diverts attention from other forms of rationality and demonstration that cut across heterogeneous discourses and exploit ambiguity. The case studies I present here, drawn from number

E.R. Grosholz, *Starry Reckoning: Reference and Analysis in Mathematics and Cosmology*, Studies in Applied Philosophy, Epistemology and Rational Ethics 30, DOI 10.1007/978-3-319-46690-3_1

theory, and the study of time and cosmology, support my claim that innovative mathematical and scientific discourse is typically heterogeneous, involving a variety of diverse idioms and notations that are superposed and juxtaposed on the page; and that the deeper epistemological explanation of this fact is that analysis and reference are disparate requirements for the inquirer. The mismatch between reference and analysis is at once productive and problematic for the advance of knowledge.

Because human awareness is both receptive and active, an accommodating construal and an explanatory construction, I warn against two kinds of errors in contemporary philosophical epistemology, evident in debates between realists and constructivists. Some philosophers demand that true knowledge be an accurate construal of the way things are; but then they deny the obvious fact that all representation is distortion, however informative it may be, and that representation itself changes the way we refer to things. Some important objects of investigation moreover cannot be directly perceived. And explanatory analysis goes far 'beyond' the things that invoked it; innovative explanation thus often sacrifices concrete, descriptive accuracy. Other philosophers want to suppose that all knowledge, and indeed all reality, is a human construction; but then they deny the obvious fact that the world is the way it is whether we like it or not, and that it has depths that elude our construals and constructions altogether. Many an explanatory analysis has shipwrecked on the hidden shoals of reality. (For stimulating discussions of current dialectic concerning realism and idealism, see Della Rocca 2016; Wagner 2017: Chap. 3, meditations that take up the dialectic by reviving Idealism on the one hand and Pragmatism on the other, in quite innovative ways). The preferable view of human knowledge, I think, lies in between. If we see it as a combination of awareness and analytic interpretation, then we will pay more attention to how in mathematics and the mathematical sciences, we employ multiple modes of representation, registering the tension between reference and analysis, as well as doing more justice to what we are investigating. I begin with a general discussion of the epistemological insights of G.W. Leibniz and Ernst Cassirer, two philosophers whose insights have particularly inspired me as I worked through my case studies during the past decade.

1 Leibniz, Locke and Analysis

For Leibniz, the key to understanding scientific knowledge is to start with the propositional schema 'S is P'. When claims of the form 'S is P' emerge in mathematical and scientific discourse, the term S typically names a problematic object like π, or the transcendental curve the tractrix, or the solar system. That is, it can be used to refer to something that is well enough understood and stable enough as an existing object to figure in a well-defined problem, when we think we are in a good position to find out more about it. Finding out more about a problematic object is, for Leibniz, a process of analysis, a search for the conditions of intelligibility of a thing, what he would call its requisites, as well as a search for the conditions of solvability of the problem which involves the thing under scrutiny. For reasons that I discuss below, analysis typically re-locates the problematic object in a relational network, where its analogies to other, similar things can be explored.

If, inspired by Leibniz, we take analysis as the key to understanding mathematical and scientific knowledge, we should also reclaim the intensional interpretation of the concept, and reaffirm the importance of non-deductive modes of reasoning, including abduction and IBE (inference to the best explanation), reasoning to general methods from particular solved problems, induction, and reasoning by analogy. The problematic object is 'there', that is, it is present or presented—we are aware of it; but analysis typically inserts the problematic object into relational networks of 'what is not there'. In the *New Essays on Human Understanding*, Leibniz argues correctly that Locke's insistence on the priority of sense perception, his extensional treatment of concepts, his Aristotelian account of abstraction, and his dismissal of innate ideas impair his account of knowledge, specifically the process of analysis (Leibniz 1962, V: 39–506, 1982). I also want to add the critical reflection that too heavy an emphasis on analysis leads to an account of scientific knowledge that slights reference, the public and clear acknowledgment of what we are talking about, as we will see in the case of Cassirer.

If we arrive at the conclusion that epistemology needs to combine versions of Leibnizian rationalism and Lockean empiricism, we may conclude further that human knowing is a hybrid process, which results in texts that are also hybrid, that is, where disparate registers (those that help us to refer successfully, and those that help us to analyze well) combine in ways that violate the assumptions of formal logic. Thus, while formal logic can help us understand certain features of mathematical and scientific rationality in historical texts, those that are captured by the forms of deductive inference, there are other features to which it must remain blind. Even the extension of logic to include formal semantics is limited by assumptions of discursive homogeneity, and the extensional interpretation of concepts. What we need is a robust account of knowledge that accounts for the 'thereness' of problematic objects, as well as the need to acknowledge the irreducible involvement of 'what is not there' when we offer mathematical proofs and scientific explanations.

Two common philosophical terms, 'abstraction' and 'instantiation', are often invoked by philosophers to finesse the disparity between these registers of language and representation in scientific texts, as if it were a simple matter to reconcile reference and analysis. On the contrary, this reconciliation takes a great deal of hard work, and its challenges engender a wide variety of strategies. At this point, it will be useful to revisit some of Leibniz's remarks on abstraction and the related notions of nominal and real definition, as well as remarks on analysis. If we understand abstraction in terms of Leibnizian analysis, we discover that it must be ampliative; because it adds content, the kinds of reasoning involved cannot be simply deductive. Ernst Cassirer makes this insight the centerpiece of his book *Substance and Function* (Cassirer 1910/1980, 1923/1953). However, from this clarification of the importance of analysis, we cannot infer that analysis supplants or replaces reference, or diminishes the hard work of bringing reference and analysis into rational alignment; and this is a point that Cassirer, in his enthusiasm for the conceptual power of functionalization and systematization, does not adequately acknowledge.

John Locke, in *An Essay Concerning Human Understanding*, begins his account of abstraction in Book III, Of Words, Chap. III, Of General Terms (Locke 1959, II: 14–31). He begins with a declaration of his nominalism by stating that all things that exist are particulars, and yet most words are general terms. Because it would be impossible and useless for every particular thing to have a name, such a language would not contribute to the sharing and improvement of knowledge. But how are general words created? His account is genetic, given in terms of the language acquisition of children. We begin with particular ideas, 'well-framed' and representing only individuals, like 'Nurse' and 'Mama'.

> Afterwards, when time and a larger acquaintance have made them observe that there are a great many other things in the world, that in some common agreements of shape, and several other qualities, resemble their father and mother, and those persons they have been used to, they frame an idea, which they find those many particulars do partake in; and to that they give, with others, the name 'man,' for example. And thus they come to have a general name, and a general idea. (Locke 1959, II: 17–18)

Thus there is nothing new or added in the general idea 'man'; it omits what is peculiar to the child's idea of each familiar person and retains only what they have in common. General ideas are thus "abstract and partial ideas of more complex ones, taken at first from particular existences", simpler and with less content. And Locke concludes, "this whole mystery of genera and species… is nothing else but abstract ideas, more or less comprehensive, with names annexed to them" (Locke 1959, II: 18–19). We should be careful not to accord abstract ideas any ontological significance, as suggested by the scholastic term 'essence': "*general* and *universal* belong not to the real existence of things, but are the inventions and creatures of the understanding, made by it for its own use, and concern only signs, whether words or ideas" (Locke 1959, II: 21). The nominalist approach thus suggests that the best way to think about a general concept is extensionally: since it is just a convenient sign applicable indifferently to many particular things, that collection of particular things is ontologically all that it amounts to, the sum total of what it really means.

Any teacher of introductory logic knows that if for the sake of pedagogy you want to refer to a collection of objects whose unity, existence and countability are relatively unproblematic, it is convenient to choose people (who can be counted by a census and identified by proper names) or natural numbers (which can be counted by school children and labeled by Indo-Arabic numerals). By offering the genetic story of abstraction in terms of a child's familiar people and the general term 'man', Locke has bracketed a number of important issues that arise for scientists as they struggle to find clear and public means for indicating what they are talking about. This struggle involves instrumentation, notation, and theoretical debate. What entities count as sufficiently unified, organized, irreducible, and stable to function as referents in a scientific discourse? Does nature reveal them to us, must we discover what nature is hiding, do we decide by arbitrary convention or by rational convention? Do scientific methods that establish and enhance reference change with historical epoch? Do theories that underwrite taxonomies change with historical epoch? How do we assess and count things with indistinct boundaries, or things that seem to be infinite? How do we assess and count things that are radically dependent on other things for their very existence? How do we assess and count the things that count things? What ontological status does science confer on the (relatively unified, organized, irreducible and stable) objects we perceive with our human sense organs, and does that status change over time? In recent books that address these questions historically, the importance of perceptual evidence (including the results of computation) to the scientific aim of good reference is clear (see Bertoloni-Meli 2006; Daston and Galison 2010; Klein and Lefèvre 2007; Azzouni 2000). So Locke's insistence that all knowledge begins in sense perception seems promising here; but by choosing relatively unproblematic illustrations, Locke doesn't confront the issue of reference as well as he might.

Probing the limitations of Locke's account of abstraction in the *New Essays*, Leibniz reveals the philosophical power of his account of analysis as the search for conditions of intelligibility. Leibniz is a nominalist in the sense that he believes that everything that has been actualized by God in this created world is an individual, that is, an irreducible, intelligible existent. However, he also thinks that different kinds of things (numbers, circles, stars, plants, people, unactualized possibles) exist in different ways. He also thinks that *phenomena bene fundata* are half-real, not mere arbitrary constructions of human reason, not mere *entia rationis* or *compendia loquendi*. One of the roles of analysis is to find the basis (among substances) for *phenomena bene fundata* that lends them reality.

Locke treats abstract terms as shorthand names that allow us to refer by one word to many individual things; we arrive at these names by 'framing ideas' that include resemblances and omit differences. For Locke, abstraction pertains not only to the ascent from species to genus, but also to the ascent from ideas of 'particular existences' to species, "by separating from them, by abstraction, time and place and any other circumstances that may determine them to this or that particular existence". For Leibniz, however, this Lockean account is plausible only for 'nominal' definitions, and for the ascent from species to genus; it suffices for the merchant or farmer. But philosophically responsible knowledge requires 'real' definitions, a knowledge of how the problematic object is produced and evidence that it really exists (Leibniz 1962, V: 268–274).

In mathematics, this requirement means that we must furnish a constructive proof of the existence of the thing, and show that its features devolve from its construction. We can usually do this in a finite number of steps, so the proof can be deductive. In science, it means that we must investigate how the problematic object is generated, and show how its features devolve from a process of generation that settles into stability. In history, it means that all cultural institutions, including language itself, have genealogies whose investigation will explain their variety, stability, and interconnection. Thus demonstrations in natural science and history must be non-deductive, because in the created world "individuality involves infinity" (Leibniz 1962, V: 268). For Leibniz, to exist as a unified thing is to be an ensouled body (composed of ensouled bodies), which is thus sentient and expressive of the world in which it finds itself; its infinity is both corporeal and psychical. So too the process of analysis must remain open-ended.

Locke's nominalism suggests an extensional account of concepts, whereas Leibniz's nominalism requires an intensional account of concepts. Locke assumes that we are aware of the problematic objects we must investigate, but doubts that we can ever know the 'secret springs' that produce them. The best methodological advice Locke can give is to guard against dogmatism; we should always be prepared to revise our stock of 'nominal essences', which are best understood as convenient labels for sets of objects. By contrast, Leibniz counsels us to look for the secret springs, even though we may never penetrate to their ultimate, infinitary source; thus we should seek intensional, real and not merely nominal definitions that explain how objects come to be what they are and why they stand in systematic, not accidental, relation to each other.

Leibniz's question for Locke is then, how do we know which resemblances are important and which differences to ignore? This question can't be answered unless we know something about the ground or cause of the resemblance of things, and the reason why certain traits go together and others do not. Abstract universal terms are rational because they are not arbitrary; as possibilities, they exist independent of our thought. Reason allows us to discern the asymmetrical relations among concepts: some concepts are presupposed by other concepts, while some concepts are incompatible with each other. And it is in virtue of this discernment that we are able then to assess the resemblances among things. General and abstract ideas are not the *result* of the perception, comparison and classification of objects but are what make that perception, comparison and classification possible.

For Leibniz, to escape nothingness is to be intelligible, that is, to exist in the way that possibles exist; to be created is to be implicated in, involved in the network of the most perfect, indeed self-perfecting, possibles. So in this world we find a kind of enhanced intelligibility. In Sections VI and VII of the *Discourse on Metaphysics*, Leibniz asserts that events express certain rules, and that nothing happens that is not rule governed (Leibniz 1962, IV: 427–463, 1989: 35–68). Access to these rules must come from the analysis of individual substances, or other intelligible things, and different kinds of things require different methods of analysis. The result of analysis is the finite expression of truth, when a predicate is asserted truly of a subject, that is, when we assert truly that something is a condition of intelligibility for something else (Leibniz 1962, IV: 431–432).

Thus Sect. VIII of the *Discourse on Metaphysics* dispels skeptical anxiety by introducing the *in-esse* theory of truth, in a statement that begins by rejecting the Lockean assumption that the predicative relation is a concatenation of disparate elements, or that it is the imposition of a 'concept' on the 'object' (Leibniz 1962, IV: 432–433). It is rather an expression of the subject. Leibniz puts the emphasis on the unity—and therefore its existence as an intelligible thing—of the subject: the predicative relation is an explicating or rendering-explicit of what the subject involves. Study of the *Generales Inquisitiones* in tandem with the *Discourse on Metaphysics* reveals an attempt to develop a theory of truth based on the assumption that the fundamental structure of reality is monadological. An intelligible thing (unified and existent in virtue of its intelligibility) is what is known, and its unity is always prior to its analysis: the unity of mind is prior to the multiplicity of things known, the transcendental number π is prior to the sequence of its digits, the unity of the organism is prior to its parts, and so forth. And so too discourse is finite: it precipitates assertions like "S is P". Yet given the nature of analysis and the things analyzed, it is also open-ended; the discovery of conditions of intelligibility does not close off inquiry, but sets the stage for further, deeper inquiry as things are better and better articulated. Discourse grows as sequences and networks that express the intelligibility of things: thus science is born, and mathematics, and history (see Grosholz and Yakira 1998: Chap. 2).

To answer the question, what the mind must be like so that it finds the world intelligible, Leibniz performs an analysis: he leads the reader back to the simple term of the individual substance. Leibniz knows he must also answer the converse question: what must the world be like so that the mind finds it intelligible? The simplex must be re-embedded in its complex surround, that is, the other *points de vue* of other individual substances. Leibniz's doctrine of pre-established harmony makes clear why analysis whose schematic outcome is "S is P" may and must be elaborated into networks of relational representations of the kind Leibniz called characteristics. The pre-established harmony is the metaphysics that underwrites the movement from the subject-predicate schema of the term-judgment to relational characteristics, like algebra or the infinitesimal calculus, which apply ambiguously to different kinds of objects and operations; conversely, we know that the pre-established harmony is the correct metaphysics because the spectrum of characteristics is the instrument we use to investigate, ever more successfully, the intelligible things we encounter (see Breger 2016: Chap. 7).

Analysis is what we use to discover what is primitive, or fundamental, or 'simple' in the investigation of a complex thing; and this investigation, in making the implicit explicit, uncovers the general or canonical form of the thing, the general formula according to which it can be treated systematically. The result is that many different cases can be expressed in terms of a single rule, equation, construction, et cetera; and we are on our way to an elaborated characteristic. Leibniz often refers to this expression of many cases by a given formulation as a 'harmony': the way that a characteristic organizes disparate cases is by analogy, not identity, as various passages from the *Discourse on Metaphysics* suggest. The establishment of the elements of a characteristic follows upon a long and difficult process of analysis; they do not, as Descartes assumes, leap to the eye (see Grosholz 2001).

Leibniz knows very well, and often remarks when describing the use of characteristics in the *ars inveniendi*, that writing allows us to say more than we know we are saying, and the best characteristics magnify this generative power, especially when they ambiguously refer to more than one kind of subject matter. A good characteristic advances knowledge not only by choosing especially deep fundamental concepts but also by allowing us to explore the analogies among disparate things. Thus he explains to Huygens that his *analysis situs* will allow him to investigate not only geometric objects but machines as well (Leibniz 1978, II: 17–25). And the later thrust of his study of differential equations is driven by his conviction that they are the key not only to understanding higher curves but also to understanding mechanical situations in terms of force.

2 Cassirer and the Rule of the Series

Ernst Cassirer is one of the great 20th c. philosophers indebted to Leibniz's central insights (see Rabouin 2009: Annex II; Grosholz 2014: 97–108). His first book, published in 1902, was devoted to Leibniz: *Leibniz' System in seinen wissenschaftlichen Grundlagen* was the first in a series of books wherein Cassirer considers the important consequences of the application of mathematics to nature for European culture and metaphysics (Cassirer 1902). The first chapter of Cassirer's *Substanzbegriff und Funktionsbegriff* is entitled, "On the Theory of the Formation of Concepts", where his treatment of abstraction owes a great deal to Leibniz (Cassirer 1910/1980: 3–34, 1923/1953: 3–26). He begins by noting the modern transformation of logic, and in order to exhibit the profundity of that change, he returns to Aristotle. Aristotelian logic, he observes, is a reflection of Aristotelian metaphysics based on substance and attribute; the logical doctrine of the construction of the concept depends on the belief that we can discover the real essences of things. The general patterns we discover in plants and animals signify both the end (*telos*) toward which the species strives, and the immanent force or *potentia* by which its evolution is guided. Only in a fixed, given, thing-like substratum can logical structure find its real application; quantity and quality, space and time, and especially relation have a strictly subordinate position in Aristotle's metaphysics, and thus his logic is primarily a logic of terms or concepts.

Aristotelian logic, Cassirer argues, generates the original problem of abstraction. If we are presented with things in their inexhaustible multiplicity and complexity, our task is to select those features that are common to several of them; things characterized by possession of some common properties are collected into classes, and this process can be reiterated to form sub-species, species, genera, and so forth. Thus the concept is a selection from what is immediately presented by sensuous reality; and every series of comparable objects has a supreme genus consisting of all the properties in which those objects agree and eliminating all the properties in which they do not agree. As we go up the hierarchy of concepts, then, the content of the more and more generic concepts diminishes, and the category of Being seems to have no content at all. Since we hope that generic scientific and mathematical concepts will give us greater and more precise determinations, this is a problem; also, logic here offers no way to distinguish between the identification of common properties that are trivial and those that are scientifically or mathematically meaningful and useful. So this account of abstraction as it is used in mathematical and scientific practice (even the practice of ancient Greece) must be incomplete (Cassirer 1910/1980: 3–11, 1923/1953: 3–9).

We have seen the same dilemma in Locke's account of general terms, compounded by his doubt that we can ever know the real essences of things. As Cassirer points out, medieval nominalism and Locke's and Berkeley's nominalist, psychological re-fashioning of the concept does not really challenge the traditional Aristotelian schema. Whereas for Aristotle external things are compared and common elements are selected from them, for Locke and Berkeley we find the same process but with presentations, psychical correlates of things, substituted for things; in this process, no new structure is produced, but only a certain division of presentations already given. John Stuart Mill continues in this vein, though since he is also a great logician, his reflections bring out its limitations. Mill tries to explain mathematical truths, for example, as abstraction from concrete matters of fact: the true positive being of every relation lies only in the individual members, which are bound together by it, and these members can only be given as individuals. The concept exists only as a part of a concrete presentation; then even arithmetic and geometry consist of propositions about certain groups of presentations. However, Mill introduces further qualifications into his account: the reproduced presentation or memory image in mathematics is somehow especially accurate and trustworthy, 'clear and distinct'. Moreover, to infer new mathematical truths, we don't need to go back every time to impressions of physical objects, because the memory-image is able to stand in for the sensible object (Cassirer 1910/1980: 11–18, 1923/1953: 9–14).

Yet the image I carry in my mind's eye of the edge of my desk is no closer to the Euclidean line than my present perception of the edge of my desk, as I type this essay. Thus it seems after all we are not concerned with concrete fact but with hypothetical forms; as Mill's own explanation shows, in mathematical concept formation the world of sensible presentations is not so much reproduced as supplanted by an order of another kind. And this observation holds equally, according to Cassirer, for theoretical physics. Thus the empiricist, psychological account of concept formation is just as problematic as the Aristotelian: similarities among things are supposed to imprint themselves on our mind while individual differences fade away. But how are we to understand this act of identification which is alleged to be the foundation of abstraction?

The act of thought, which holds together two temporally separate presentations, and recognizes them as in some way related, does not itself have an immediate sensible correlate in the contents compared, Cassirer argues. Aristotle, Locke, Berkeley and Mill all presuppose that things or psychical presentations have already been ordered into 'series of similars' for the logician's consideration. An unacknowledged, unarticulated construction of a series stands behind their accounts of the construction of concepts. This series has already, tacitly, been generated by a rule which is then manifest in the specific form of the concept. Once we see this, we can also see the important inference Cassirer draws from it: there are many kinds of 'rules of the series'. Aristotle, the medieval philosophers, and the empiricists, enchanted by the category of substance, limited themselves to series in which the connection among the members of the series is the possession of a common element or property: αa, αb, αc, αd.... But (as Leibniz often explained and showed in his infinitesimal calculus) there are many other kinds of possible connections among members of a series: we might have a, a + b, a + b + c, a + b +c + d, a + b + c + d + e, … for example, or a, b − a, c − b, d − c, e − d, …, or indeed any useful function binding the terms. Then we must see as well (as Leibniz saw) that this rule of the series is something *extra*: it is not arrived at by mere summation of terms or neglect of parts of terms (Cassirer 1910/1980: 18–23; 1923/1953: 14–18).

At this point, Cassirer comes to the notion of mathematical and scientific (physical) function which is the central idea of his book. Invoking Johann Heinrich Lambert, an 18th c. critic of Leibniz's follower Christian Wolff, he claims that mathematical concepts do not cancel or forget the determinations of the special cases, but fully retain them. When a mathematician makes his formula more general, Cassirer asserts, this means that he is able not only to retain all the more special cases, but also to deduce them from the universal formula. Note that if a general concept had been arrived at by Aristotelian abstraction, the special cases could not be recovered from it, because the particularities have been forgotten. By contrast, the mathematical or scientific concept seeks to explain and clarify the whole content of the particulars by exhibiting their deeper systematic connections, revealed in the law of the series. Here the more universal concept is more, not less, rich in content; it is not a vague image or a schematic presentation, but a principle of serial order; indeed, it is the kind of intensional, 'real' definition that Leibniz advocated for mathematical research. Thus in modern mathematics, things and problems are not isolated, but shown to exist in thoroughgoing interconnection (Cassirer 1910/1980: 24–34, 1923/1953: 18–26).

Cassirer argues that the abstract concept emerges not from disconnected particularities, but from elements that have already been presupposed as organized in an ordered manifold. The rule of the series is not arrived at through bare summation or neglect of parts. The real problem of abstraction, he concludes, is to identify the thorough-going rule of succession that has been presupposed, and to articulate it. Using Descartes' reorganization of geometry by means of algebra as an example, Cassirer shows that the explicit articulation of functions and relations in modern mathematics allows us to deploy general concepts that do not cancel the determinations of the special cases, but in all strictness fully retain them, and moreover reveal deeper systematic connections among instances formerly regarded as disconnected. Abstraction thus understood increases content.

Furthermore, Cassirer continues, the concrete universality of the mathematical function (despite the way in which Hegel opposes it to the abstract universality of the concept) extends to the scientific treatment of nature. Thus a series of things with attributes is transformed into a systematic totality of variable terms or parameters; things are transformed into the solutions of complex equations, as when a molecule becomes the solution to a wave equation, or when the sun, the moon and the earth become a solution to the three-body problem. As Cassirer wrote earlier, the world of sensible presentations is not so much reproduced as supplanted by an order of another kind. Then whenever we unify the 'first order' objects of our thought into a single system, we create new 'second order' objects whose total content is expressed in the functional relations holding among them. So too, no summation of individual 'first order' cases can ever produce the specific unity which is meant by the system unified under the functional concept. Cassirer insists on the distinction between the members of the series and the form of the series: "the content of the concept cannot be dissolved into the elements of its extension… The meaning of the law that connects the individual members is not to be exhausted by the enumeration of any number of instances of the law; for such enumeration lacks the generating principle that enables us to connect the individual members into a functional whole" (Cassirer 1910/1980: 33–34, 1923/1953: 26). Leibniz would have approved this claim.

Cassirer in *Substanzbegriff und Funktionsbegriff* has identified a supremely important conceptual innovation, which characterizes to a great extent the mathematical and scientific thinking that we like to call modern. In this kind of conceptualization, abstraction or the ascent to a more general concept does not diminish content, but rather increases it. Such ampliative abstraction both conserves the content of the particulars and adds to them a representation of their systematic interconnection, which then moreover may promote the annexation of ideal elements and generate further knowledge. Thus analysis rewrites geometry as analytic geometry, and allows Leibniz and his followers to annex the transcendental functions and infinites series; the Peano Postulates and Dedekind's set theory rewrite arithmetic and allow Cantor to annex the transfinite numbers; and Frege and Russell rewrite logic and allow the annexation of the hierarchy of classes of recursive sets.

3 A Criticism of Cassirer: What's in a Circle?

All the same, Cassirer is a bit hasty and optimistic when he proclaims that the mathematical concepts that result from such ampliative abstraction, governed by a search for the 'rule of the series', do not forget the determinations of the special cases, but fully or wholly retain them: that is, the mathematician is able not only to retain all the more special cases, but also to *deduce* them from the universal formula. Then the work of analysis could be substituted for the work of reference. However, this claim is in fact inconsistent with Cassirer's further claim that geometry (and physics) must be transformed in order to be subject to such serial systematization: the world of sensible presentations is not so much reproduced as supplanted by an order of another kind (Cassirer 1910/1980: 88–147, 148–310, 1923/1953: 68–111, 112–233). He insists on this radical transformation of the special cases, because it is very important to his attack on Aristotelianism that the thought process of ampliative abstraction is not inductive.

In analytic geometry, for example, the circles, ellipses, parabolas and hyperbolas that we find in Greek geometry must be transformed into algebraic equations in order to be comprehensible as specifications of a single general equation that exhibits their systematic interconnection. However, those geometric forms, represented by diagrams and investigated by the Greek geometers, by Leibniz, by the projective geometers and by Hilbert, present information that is not included in the algebraic equation. Even though the algebraic equation manages to express a great deal of information about, for example, the circle, the content of the circle (that is, the role it plays in an on-going series of proposed and solved problems) is not 'wholly' or 'fully' represented by the algebraic equation. The content of the circle, for any 21st century student of mathematics, includes the way in which it embodies as a kind of compendium the sine and cosine functions; but this content is not represented by the algebraic equation. Thus, first of all, if the members of the series have been radically transformed by the process of ampliative abstraction, they themselves cannot be deduced from the rule of the series: apples cannot be deduced from oranges.

But then what is a circle, that stubborn referent? Suppose that our knowledge of the circle consisted in unpacking "what was already contained in the definition", which is what Kant asserts. Euclid's definition runs: "A circle is a plane figure contained by one line such that all the straight lines falling upon it from one point among those lying within the figure are equal to one another, and the point is called the center of the circle". This is a good definition, soon thereafter followed (right after the postulates and common notions) by Proposition I, which presents a diagram with two circles, used to construct an equilateral triangle on a given line segment (Euclid 1956: 153 and 241). There is a scholarly and philosophical debate about how important the diagram is in relation to the definition, related to a dispute about at what point historically diagrams became part of the Euclidean tradition (see Netz 2003).

We cannot deduce from Euclid's definition the fact that the sine and cosine functions can be generated in terms of the circle. They are 'contained in' the concept of the circle for Euler, but not for Euclid. How did they show up there, right in the middle, plain as day? The answer is given by Leibniz's notion of analysis, for Descartes and Leibniz himself discovered new conditions of intelligibility for the circle by bringing it into novel relation with algebra and arithmetic. We need the Pythagorean theorem (which is about triangles), Cartesian geometry (which is not only about geometry but also about the algebra of arithmetic), the infinitesimal calculus, the completion of the rationals that we call the reals and the notion of a transcendental function. Once we have embedded the circle in Cartesian geometry, algebra, arithmetic and infinitesimal analysis, and Euler has developed the modern notion of a function for us, we can see that the functions sine and cosine can be 'read off' the circle (see Youschkevitch 1976).

We impose two Cartesian coordinates on the circle, intersecting at its center, in such a way that, normalizing, we set the length of the radius of the circle $r = 1$ (Fig. 1). We then examine the invariant relations among the sides of the right triangles—each with its base on the x-axis, one vertex on $(0, 0)$ and one vertex on a point of the circle (x, y)—generated as the radius turns and the angle, θ, increases. Then we see that the changing y-coordinate yields the sine function and the changing x-coordinate yields the cosine function, and so, by the Pythagorean Theorem, $\sin^2\theta + \cos^2\theta = 1^2 = 1$, a source of never-ending amazement to 17th c. and 18th c. mathematicians.

Fig. 1 Sine and cosine functions in the *circle*

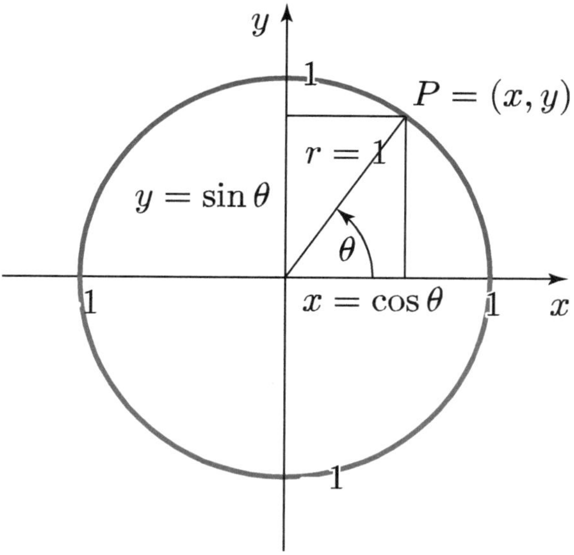

The embedding of the circle in the emerging field of real analysis required mathematicians to lift the circle into its new context and then to regain it, or reaffirm it, as the object it was to begin with. The re-construction of the unit circle 'upstairs' in real analysis as the locus of the equation $x^2 + y^2 = 1$ plays an important role in the choice of notation and procedures, and helps to provide a conceptual grounding for the introduction of the concept of algebraic curves and a bit later, by contrast, that of transcendental curves. The reinsertion of the circle-equation back into geometry as the original circle provides an important conceptual check and guide for a wide variety of developments, and also changes the 18th century understanding of geometry in ways that are then only fully recognized in the early 19th century. The enterprise of lifting and reinsertion, pertinent to many other kinds of things besides circles, is an important pattern of generalization in tandem with re-particularization that precipitates many of the procedures and methods characteristic of the beginnings of real analysis. In Descartes' *Geometry*, Descartes never launches into his stunningly ampliative, algebraic reasoning without constantly checking the cogency of his procedures and novel items back against geometry. The same might be said of the quite ampliative exposition of Cartesian geometry found in Franz van Schooten's *Geometry by René Descartes* (1659–61), which leads directly to the development of the infinitesimal calculus in the hands of Huygens and Leibniz, and Barrow and Newton (see Bos 2013).

The 19th c. projective geometer Michel Chasles criticized the Greek geometers, observing that the lack of generality in proofs was a real drawback for the Greeks, and prevented them from making progress. Each construction used to determine the tangent to a curve, for example, depended on specific features of the curve and therefore used procedures specific to the object being studied; each method, he lamented, was "essentially different". There was one method used by the Greeks that seemed general, the 'method of exhaustion', but Chasles argued that the various ways in which the method was applied were not general (Chasles 1889). If we focus on the 'modalities of application' (Karine Chemla's useful phrase) of general methods to which Chasles draws our attention, we see that only in the 17th century were uniform means provided for the application of a general method for determining the tangent to a curve (Chemla 1998). Chasles points to Descartes' method of tangents (analytic geometry) and Cavalieri's and Leibniz's method of indivisibles (infinitesimal analysis) as examples. He also points to Desargues' treatment of the conic sections (projective geometry). In the case of Descartes' method of tangents, there is a general method, but it can be applied to curves only when one has succeeded in representing them by an algebraic equation. That is, the modality of application requires that the curve be replaced by an equation. Obviously, this method at first wouldn't work for transcendental curves, precisely because they are not algebraic! Some of Leibniz's early efforts in developing the infinitesimal calculus are attempts to apply the method of tangents to non-algebraic curves.

The extension of the general methods of analytic geometry to transcendental curves in the long run required novel (non-Cartesian) modalities of application for the same general method. Curves are re-represented as equations with an infinite number of terms, equations with variable exponents, equations with log, e, sin and cos as basic components, and equations re-conceptualized as functions. Then functions (with real variables) are re-represented as the solutions to differential equations, as functions of a complex variable, as Fourier series, et cetera; and these pathways lead into the 19th c. debate about what a function might be (see Lakatos 1976). There is an on-going mutual adjustment of methods and modalities of application, as well as a checking of those modalities against the things they were originally designed to cover; new general methods reorganize geometry and call up new families of functions. Thus, to prove that a certain line is the tangent to the catenary at such and such a point may require the mathematician to quit Euclidean geometry and take up the modes of representation of a different domain altogether. All the same, a line is still a line, even when the curve to which it stands as a tangent is the result of a new discipline.

In the 19th century, the circle comes to harbor another, equally surprising set of things, the *n*th roots of unity, that is, the complex numbers satisfying the equation $z^n = 1$ (Fig. 2). For this discovery to take place, the mathematical community had to embed the study of the integers and the reals in the study of the complex numbers (algebraic number theory on the one hand and complex analysis on the other do this in quite distinct ways), and further recast that study in terms of the complex plane, so that the resources of Euclidean geometry, projective geometry and eventually non-Euclidean geometry could be used to frame, unify and solve clusters of problems. On the complex plane, the *n*th roots of unity are the vertices of a regular *n*-polygon inscribed in the unit circle, $|z| = 1$, where ($\sqrt{-1} = i$, $z = x + iy$ and $|z| = x^2 + y^2$. We can also 'read off' the unit circle, understood this way, as the set of points (x, y) where $x = \cos\theta$ and also $y = \sin\theta$. Then Euler's famous

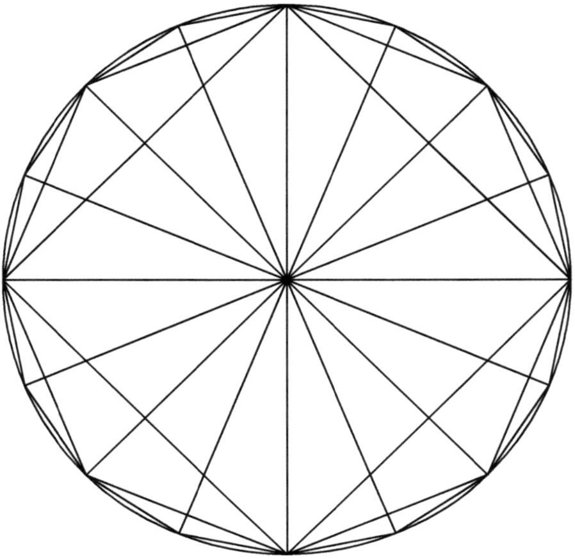

Fig. 2 *Circle* with fourth, eighth and sixteenth roots of unity

formula, $e^{i\theta} = \cos\theta + i\sin\theta$, is valid for all real θ, along with the representation of the nth roots of unity ζ_n as $e^{2\pi i\theta k/n}$ (see Nahim 2010; Mazur 2004).

The field **Q**[i] is an algebraic extension of the field of rational numbers **Q**, obtained by adjoining the square root of -1 ($\sqrt{-1} = i$) to the rationals. Its elements are of the form $a + bi$ where a and b are rational numbers, which can be added, subtracted and multiplied according to the usual rules of arithmetic, augmented by the equation $i^2 = -1$; and a bit of computation shows that every element $a + bi$ is invertible. Within this field **Q**[i] we can locate the analogue of the integers **Z** within **Q**: it is **Z**[i], the Gaussian integers, whose elements are of the form $a + bi$ where a and b are integers. Like **Z**, **Z**[i] enjoys unique factorization (that is, each nonzero element of **Z**[i] can be expressed as a unique (up to units) product of primes), but its primes are different from those of **Z**; and instead of two units it has four: 1, i, -1, and $-i$. To be a prime in **Z**[i], $a + bi$ must satisfy these conditions: either a is 0 and b is a prime in **Z** and is congruent to 3(mod 4); or b is 0 and a is a prime in **Z** and is congruent to 3(mod 4); or neither a nor b are 0 and $a^2 + b^2$ is a prime in **Z** and is not congruent to 3(mod 4). When we use the Euclidean plane as a model for **C**, the units are then modeled by the square with endpoints 1, i, -1 and $-i$. This suggests the generalization that models the set of nth roots of unity as vertices of regular n-polygons centered at 0 on the complex plane, with one vertex at 1. This nesting of all such polygons within the circle on the complex plane provides a kind of visual and iconic index for the generalization from **Q**[i] to other algebraic fields called cyclotomic fields, where an nth root of unity (ζ_n) is adjoined to **Q** (**Q**[i] is **Q**[ζ_4]). Those roots of unity, regarded as vertices, suggest the notion of symmetry, and in that light may be studied in terms of groups of symmetries. For each cyclotomic field $\mathbf{Q}[\zeta_n]$ there is a group of automorphisms that permutes the roots of unity while mapping $\mathbf{Q}$ to itself; it is called the Galois group Gal $(\mathbf{Q}[\zeta_n]/\mathbf{Q})$ (Kato et al. 1998: Chap. 5).

So I would argue that the circle becomes 'problematic' in the context of Euclid's geometry, in relation to the lines and triangles that can be inscribed in it or circumscribed around it; many of these problems are resolved. It becomes 'problematic' again, in different ways and in relation to other kinds of things, like polynomials and transcendental curves in the late 17th century; many of those problems are resolved. Then in the 19th century the circle emerges again, 'problematic' with respect to newly constituted things, like the complex plane, holomorphic and diffeomorphic functions, cyclotomic fields and more generally algebraic number fields. That is, a mathematical concept like the circle is not problematic in isolation but in relation; and its problematicity waxes and wanes. This kind of generalization brings new content to the circle by putting it in novel relation to novel things. The reason the novel juxtapositions I have just sketched are fruitful for mathematics is because the circle is what it is and has an irreducible unity. In all of these contexts, the circle severely constrains what can be said of it and how it can be used to frame and encompass other things. The precise and determinate resistance the circle offers to any use made of it contributes to the growth of knowledge. The circle proves itself again and again as a canonical object. The work of analysis must always be supplemented by, and exists in tension with, the work of reference (see Magnani 2001: Chaps. 6 and 7).

Thus we cannot simply identify the circle with, say, the equation $x^2 + y^2 = 1$; for the reasons just given, this representation of the circle does not wholly retain all the 'determinations' of the circle or all of its content. (Moreover, considered over fields or rings different from the reals, it does not in fact represent a circle.) For the circle to reveal the important 'content' of its relation to the sine and cosine functions, for example, it had to be set in the context of 18th c. analysis, re-inscribed in terms of a very different 'rule of the series', and thus represented in a novel way. Indeed, in his discussion of analytic geometry and projective geometry, Cassirer acknowledges that the things of geometry can be investigated by two quite different kinds of ampliative abstraction, governed by different serial rules, and then can be made to yield rather different swathes of mathematical knowledge (Cassirer 1910/1980: 99–119, 1923/1953: 76–91).

And this point also counts against the structuralism which he too enthusiastically endorses when he writes that the aim of the act of ampliative abstraction is to bring to consciousness the meaning of a certain relation independently of all particular cases of application, purely in itself, or that ampliative abstraction results in a system of ideal objects whose whole content is exhausted in their mutual relations. Structuralism can't explain how we identify instances across logically incongruent systematizations, or why we would be motivated to do so. How does Cassirer know to identify the circle in Euclidean geometry, in analytic geometry, and in projective geometry, unless he is dealing with a mathematical thing that asserts its independent, rather Aristotelian and substance-like, existence as it lends itself to different modes of representation over the ages and across disciplines? Indeed, who knows which mode will turn up next, and what we will learn next about the circle?

Cassirer is most structuralist in the second chapter of *Substanzbegriff und Funktionsbegriff*, when he discusses number, and its treatment by Dedekind on the one hand and by Russell on the other. He seems to think that the 'primitive' view of numbers, revived by Pythagoras and Plato in one way and by "Mill's arithmetic of pebbles and gingerbread nuts" in another, is an illusion. Numbers cannot be independent existences present anterior to any relation, or 'given' in isolation from each other. Of numbers, he says that the transformation that we have noted in the case of the serial systematization of geometrical objects is merely copying, which does not produce a new thing. To subject numbers to the rule of the series is just to represent them as they really are, mere placeholders in a relational structure, and to banish the misguided way in which we might earlier have thought about them (Cassirer 1910/1980: 35–87, 1923/1953: 27–67).

But different rules of the series bring out different features of numbers, and help to solve different families of problems about them, as I will show in the chapters that follow. How is Cassirer able to re-identify the number one, or the number one million million, or the sequence of prime numbers, or the sequence of Fibonacci numbers, from one context to another when they have been ampliatively abstracted in different ways? Andrew Wiles proved Fermat's Last Theorem by setting the natural numbers in a series of such contexts, which have a limited relation to the context provided by the Peano Postulates, beloved of logicians. On the other hand, number theorists seem to have little to say about whether the natural numbers, as they continue into the transfinite, pursue a linear course or become rather more tree-like, branching as they go. This is a compelling and important question, but number theorists don't seem particularly interested in going beyond ω. Number theorists on the one hand, and set theorists on the other, approach the natural numbers with very different serial systematizations; yet each group knows that it is concerned with the same mathematical things that concern the other. How do they know that?

Ampliative abstraction, which acknowledges that it investigates things in series and strives to articulate the rule of the series, does manage to conserve much of the content of the particulars, as it exhibits the systematic interconnections that bind them, thus adding rather than subtracting content in the general concept. Yet it does not 'wholly' exhaust the content of the particulars, and it does not reduce the particulars to mere nodes in a net of relations, not even in arithmetic. It does not support the structuralism that seems to inspire Cassirer, and it does not allow us to escape the task of reference. Biology, chemistry, physics, geometry and arithmetic cannot in principle, and in practice never do, dispense with the things that occur in the problems that characterize them, and that shape their methods. Leibniz's metaphysical doctrine of the monad, and in a different sense his allegiance to geometry, as well as his anti-reductionist disposition, save him from the structuralism that tempted Cassirer.

References

Azzouni, J. (2000). *Knowledge and reference in empirical science*. New York and London: Routledge.

Bertoloni-Meli, D. (2006). *Thinking with objects: The transformation of mechanics in the seventeeth century*. Baltimore: Johns Hopkins University Press.

Bos, H. J. M. (2013). *Redefining geometrical exactness: Descartes' transformation of the early modern concept of construction*. Stuttgart: Springer.

Breger, H. (2016). *Kontinuum, Analysis, Informales—Beiträge zur Mathematik und Philosophie von Leibniz*. W. Li (Ed.). Berlin: Springer.

Cassirer, E. (1902). *Leibniz' System in seinem wissentschaftlichen Grundlagen*. Marburg: N. G. Elwert'sche Verlagsbuchhandlung.

Cassirer, E. (1910/1980). *Substanzbegriff und Funktionsbegriff: Untersuchungen über die Grundfragen der Erkenntniskritik*. Berlin: Bruno Cassirer; repr. Darmstadt: Wissenschaftliche Buchgesellschaft.

Cassirer, E. (1923/1953). *Substance and function and Einstein's theory of relativity* (W. C. Swabey & M. C. Swabey, Trans.). Chicago: Open Court.

Chasles, M. (1889). *Aperçu historique sur l'origine et le développement des methods en géométrie*. Paris: Gauthier-Villars.

Chemla, K. (1998). Lazare Carnot et la généralité en géométrie. In *Revue d'histoire des mathématiques, IV*, 163–190.

Daston, L., & Galison, P. (2010). *Objectivity*. Cambridge, MA: MIT Press.

Della Rocca, M. (2016). Meaning, the history of philosophy, and analytical philosophy: A Parmenidean ascent. In M. van Ackeren (Ed.), *Philosophy and the Historical Perspective. Proceedings of the British Academy*. Oxford: Oxford University Press. Forthcoming.

Euclid. (1956). *Elements*. T. L. Heath (Ed.). New York: Dover.

Grosholz, E. (2001). Theomorphic expression in Leibniz's *Discourse on Metaphysics*. In G. H. R. Parkinson (Ed.), *Studia Leibnitiana*, thematic issue: Towards the *Discourse on Metaphysics*. Band 33, Heft 1, 4–18.

Grosholz, E. (2014). Leibniz, Locke and Cassirer: Abstraction and Analysis. In H. Breger & W. Chao (Eds.), *Studia Leibnitiana*, Special Issue on Analysis. Band 45, Heft 1, 97–108.

Grosholz, E., & Yakira, E. (1998). *Leibniz's science of the rational*. *Studia Leibniziana* Sonderheft 26. Stuttgart: Steiner Verlag.

Kato, K., Saito, T., & Kurokawa, N. (1998). *Number theory 2: Introduction to class field theory*. Providence, RI: American Mathematical Society.

Klein, U., & Lefèvre, W. (2007). *Materials in eighteenth-century science: A historical ontology*. Cambridge, MA: MIT Press.

Lakatos, I. (1976). *Proofs and refutations*. Cambridge: Cambridge University Press.

Leibniz, G. W. (1962). *Die Philosophischen Schriften* (Vol. I–VII). C. I. Gerhardt (Ed.). Hildesheim: Georg Olms.

Leibniz, G. W. (1978). *Mathmatischen Schriften* (Vol. I–VII). C. I. Gerhardt (Ed.). Hildesheim: Georg Olms.

Leibniz, G. W. (1982). *New essays on human understanding* (P. Remnant & J. Bennett, Trans.). Cambridge: Cambridge University Press.

Leibniz, G. W. (1989). *Philosophical essays* (R. Ariew & D. Garber, Trans.). Indianapolis: Hackett.

Locke, J. (1959). *An essay concerning human understanding* (Vol. I–II). A. C. Fraser (Ed.). New York: Dover Publications.

Magnani, L. (2001). *Philosophy and geometry: Theoretical and historical issues*. Dordrecht: Kluwer.

Mazur, B. (2004). *Imaginary numbers*. New York: Picador.

Nahim, P. (2010). *An imaginary tale*. Princeton: Princeton University Press.

Netz, R. (2003). *The shaping of deduction in Greek mathematics: A study in cognitive history*. Cambridge: Cambridge University Press.

Rabouin, D. (2009). *Mathesis Universalis, L'idée de "mathématique universelle" d'Aristote à Descartes*. Paris: Presses Universitaires de France.

Wagner, R. (2017). *Making and breaking mathematical sense: Histories and philosophies of mathematical practice*. Princeton: Princeton University Press.

Youschkevitch, A. P. (1976). The concept of function up to the middle of the 19th century. In *Archive for History of Exact Sciences, XVI*(1), 37–85.

Chapter 2
Philosophy of Mathematics and Philosophy of History

In this chapter, I argue for the historicity of mathematics, and thus for the pertinence of the study of the history of mathematics to the philosophy of mathematics. I also argue that we need to look carefully at what constitutes our philosophy of history, and the nature of historical explanation. As I hope should be evident from my arguments in the preceding chapter, this doesn't mean that I think that what we say about mathematics is contingent or empirically based. Here I invoke Plato, the great dialectician of classical antiquity. The *Meno* is a dramatic dialogue in which two analyses take place, one mathematical and one philosophical. In the first, Socrates leads the slave boy to correct, refine, and extend his intuition of a pair of geometrical diagrams by examining his presuppositions in the light of certain background knowledge (about arithmetic), as they solve a problem together. In the second, Socrates leads Meno to correct, refine, and extend his intuition about virtue as well as about method by examining his presuppositions in the light of certain background knowledge, as they engage in philosophical dialectic together. In both cases, an intelligible unity is apprehended, but imperfectly, and analysis leads to increased understanding via a search for the conditions of intelligibility (Plato 1961: 353–384). The form of dramatic dialogue expresses something essential to the process of analysis, for the dialogue may be read both as a set of arguments (in all of which *reductio ad absurdum* plays a pivotal role) that uncover the logical presuppositions of various claims, but it may also be read as a narrative, a process in time and history.

And now I invoke Aristotle, the great logician of classical antiquity. A narrative, as Aristotle tells us in the *Poetics*, has a beginning, middle, and end. The beginning introduces us to characters who act in a certain situation, one or a few of whom have a special claim on our empathy and interest; the middle introduces one or more surprising contingencies or reversals, and the skill of the storyteller lies in maintaining the continuity of the story not in spite of but by means of those discontinuities, deepening our understanding of the characters and their actions; and the end draws us along throughout the tale: how will events ultimately turn out for these characters? (Aristotle 1947: 620–667). By contrast, an argument has the structure of

E.R. Grosholz, *Starry Reckoning: Reference and Analysis in Mathematics and Cosmology*, Studies in Applied Philosophy, Epistemology and Rational Ethics 30, DOI 10.1007/978-3-319-46690-3_2

a series of premises that support, in more or less materially and formally successful ways, a conclusion. The priority of premises with respect to a conclusion is logical, not temporal or historical; an argument has no beginning or end, no before and after. The logician takes the cogency of an argument to hold atemporally: if such and such is the case, then such and such must follow, universally and necessarily. Pragmatic or rhetorical considerations may superimpose an historical dimension on the argument, but that is not the concern of the logician.

1 W. B. Gallie on History

Analysis, as I characterized it in the last chapter, is the search for conditions of the intelligibility of existing things; in mathematics, this often takes the form of solving a problem. It is an ampliative process, that increases knowledge as it proceeds. From an analysis, an argument can be reconstructed, as when Andrew Wiles finally wrote up the results of his seven-years-long search in the 108 page full dress proof in the May 1995 issue of *Annals of Mathematics* (Wiles 1995: 443–551). As we will see in the following chapter, Carlo Cellucci contrasts the analytic method of proof discovery rather starkly with the axiomatic method of justification, which reasons "downwards", deductively, from a set of fixed axioms, as well as with algorithmic problem-solving methods (see Cellucci 2013; Ippoliti 2008). He argues that the primary activity of mathematicians is not theory construction but problem solution, which proceeds by analysis, a family of rational procedures broader than logical deduction. Analysis begins with a problem to be solved, on the basis of which one formulates a hypothesis; the hypothesis is another problem which, if solved, would constitute a sufficient condition for the solution of the original problem. To address the hypothesis, however, requires making further hypotheses, and this "upwards" path of hypotheses must moreover be evaluated and developed in relation to existing mathematical knowledge, some of it available in the format of textbook exposition, and some of it available only as the incompletely articulated "know-how" of contemporary mathematicians (see Breger 2000). Indeed, some of the pertinent knowledge will remain to be forged as the pathway of hypotheses sometimes snakes between, sometimes bridges, more than one domain of mathematical research (some of which may be axiomatized and some not), or when the demands of the proof underway throw parts of existing knowledge into question.

The great scholar Gregory Vlastos made a career out of reconstructing the arguments in Platonic dialogues, abstracting them from the dramatic action. And yet Platonic analysis is also a process of enlightenment in the life of an individual as well as a culture: the slave boy and Meno (but not Anytus) are changed in some fundamental way by their encounter with Socrates. They come to understand something that they had not understood before, and their success in understanding could not have been predicted. There is no way to cause understanding, nor can virtue be caused, even by a virtuous father desperately concerned about his wastrel son, as Socrates shows at the end of the dialogue. The middle of the plot is a crisis where opinions reveal their instability and things their problematicity, and it is marked by a *reductio* argument; but it is also marked by anxiety and curiosity on the part of Socrates' interlocutors, emotions that might lead them equally to pursue or to flee the analysis. And at the end of the dialogue, even though it is aporetic, a shared illumination occurs that bestows on the reader a sense of the discoverable intelligibility of things, how knowledge unfolds. To understand an argument and to follow a story are two different things, but the reader must do both in order to appreciate a Platonic dialogue.

What does this have to do with philosophy of mathematics? The dialogue *Meno* shows that the philosophy of mathematics must stand in relation to the history of mathematics, and moreover this relationship must be undergirded by a philosophy of history that does not reduce the narrative aspect of history to the forms of argument used by logicians and natural scientists. History is primarily narrative because human action is, and therefore no physicalist-reductionist account of human action can succeed (see Danto 1965: Chap. 8). So philosophy must acknowledge and retain a narrative dimension, since it concerns processes of enlightenment, the analytic search for conditions of intelligibility. Indeed, the very notion of a problem in mathematics is historical, and this claim stems from taking as central and irreducible (for example) the narrative of Andrew Wiles' analysis that led to the proof of Fermat's Last Theorem. If mathematical problems have an irreducibly historical dimension, so too do theorems (which represent solved problems), as well as methods and systems (which represent families of solved problems): the logical articulation of a theory cannot be divorced from its origins in history. This claim does not presume to pass off mathematics as a series of contingencies, but it does indicate in a critical spirit why we should not try to totalize mathematical history in a formal theory.

W. B. Gallie begins his book *Philosophy and the Historical Understanding* by rejecting the Hegelian/Marxist doctrine of an inescapable order, a rational purposive pattern in history that might lead us to predict the end of history, when all social contradictions shall be resolved. He reviews the partial insights of Cournot, Dilthey, Rickert, and Collingwood, and then remarks upon a set of logical positivist studies on the subject of historical explanation (see Gardiner 1961). Gallie writes, "These studies are, broadly speaking, so many exercises in applied logic: their starting point is always the general idea of explanation, and they tend to present historical explanations as deviant or degenerate cases of other logically more perfect models" (Gallie 1964: 19). Later in the book, he elaborates on this point: "There has been a persistent tendency, even in the ablest writers, to present historical explanations as so many curiously weakened versions of the kind of explanation that is characteristic of the natural sciences. To speak more exactly, it is claimed or assumed that any adequate explanation must conform to the deductivist model, in which a general law or formula, applied to a particular case, is shown to require, and hence logically to explain, a result of such and such description" (Gallie 1964: 105).

By contrast, Gallie argues that history is first and foremost narrative, recountings of human actions that concern the historian and his readers; and that the ending of a story is "essentially a different kind of conclusion from that which is synonymous with 'statement proved' or 'result deduced or predicted'" (Gallie 1964: 23). A good story always contains surprises or reversals that could not have been foreseen. "But whereas for a scientist a revealed discontinuity usually suggests some failure on his part, or on the part of his principles and methods and theories, to account for that aspect of nature which he is studying, for the man who follows a story a discontinuity may mean… the promise of additional insights into the stuff a particular character is made of, into the range of action and adaptation which any character can command" (Gallie 1964: 41). Indeed, Gallie adds, this is the logical texture of everyday life, where the unforeseen constantly puts to the test our intellectual and moral resources, and where our ability to rise to the occasion must always remain in question: the insight of tragedy is that anyone can be destroyed by some unfortunate combination of events and a lapse in fortitude or sympathy.

Thus, the most important relation between events in a story is "not indeed that some earlier event necessitated a later one, but that a later event required, as its necessary condition, some earlier one. More simply, almost every incident in a story requires, as a necessary condition of its intelligibility, its acceptability, some indication of the kind of event or context which occasioned or evoked it, or, at the very least, made it possible. This relation, rather than the predictability of certain events given the occurrence of others, is the main bond of logical continuity in any story" (Gallie 1964: 26). Unless one is reading a detective novel, which is not so much a story as a puzzle in narrative clothing, following a narrative is emphatically not a kind of deduction or even induction in which successively presented bits of evidence allow the reader to predict the ending: "Following is not identical with forecasting or anticipating, even although the two processes may sometimes coincide with or assist one another" (Gallie 1964: 32). Following a story is more like the process of analysis. The intrusion of contingency into a story makes it lifelike; the way we retrospectively make sense of or redeem contingencies in our lives, as in the way we understand stories, is a good example of how we lend intelligibility to things. The command to search for conditions of intelligibility is not just a theoretical but also a practical or moral command; and we honor it not just by constructing arguments but also by telling stories.

This view of history has two important, and closely related, philosophical consequences. The first is that although historians and human beings in their attempts to come to moral maturity must pursue objectivity, *wie es eigentlich gewesen war*, in order to escape the provincialism and prejudices that mar histories, they cannot give this ideal any descriptive content: it is a regulative ideal, to borrow a Kantian term. "The historian is committed to the search for interconnectedness and is thus drawn on by an ideal demand that expresses his ideal of the whole, of the one historical world. But at the same time, because of the inevitable selectiveness of all historical thinking, it is impossible that he should ever reach… that ideal goal" (Gallie 1964: 61). There can be no such thing as the Ideal Chronicle.

The second is that the physicalist-reductionist account of human action which supposes such an Ideal Chronicle recording events as they happen, in complete and accurate detail, must be renounced, as Danto argues. There are two ways to run the thought experiment. Suppose that the chronicle is written in a language rich enough to include the ways in which historians normally pick out, characterize, and link events. (This is a generous concession to the physicalist-reductionist, who is not really entitled to it.) This language contains a whole class of descriptions that characterize agents and actions in terms of their future vicissitudes as well as words like 'causes', 'anticipates', 'begins', 'precedes', 'ends', which no historian could forego without lapsing into silence. But such descriptions and words are not available to the eyewitness of events who describes them on the edge of time at which they occur. The description of an event comes to stand in different relations to those that come after it; and the new relations in turn may point to novel ways of associating that event with contemporary and antecedent events or indeed to novel ways of construing the parts of a spatio-temporally diffused event as one event. The historian's way of choosing beginnings and endings for narratives is her prime way of indicating the significance of those events.

Thus to allow a sufficiently rich language for the Ideal Chronicle violates the original supposition of how the chronicle is to be written. So let us hold the physicalist-reductionist to his own restrictions: the chronicle must be written in terms impoverished enough to meet the stringent conditions of its writing on the edge of time. Then we find that it is reduced to an account of matter in motion, and the subject matter of history, people and their actions, has dissolved (as it dissolves in Lucretius' epic *De rerum natura*) (Lucretius 2007). The chronicle is no longer about history, and the chronicler is doing descriptive physics if he is doing anything at all. If we want to write history, an event must have a beginning, middle, and end; must be related to its past and future; and must be construed as significant.

2 Kitcher and Cavaillès on History

In short, history is not science, historical explanation is not scientific explanation, historical method is not scientific method, and human action is not (merely) an event in nature. If philosophers of science and mathematics wish to make serious use of the history of their subjects, they must take these distinctions into account; indeed, this accounting will be helpful because philosophy that takes history seriously cannot pretend to be a science. In fact, philosophy, like mathematics, is neither history nor science, but involves both narrative (like history) and argument (like science), because the process of analysis involves both; the philosophical school that tries to banish either dimension will not succeed.

Up until the last couple of decades, history has been a term conspicuously absent in Anglo-American philosophy of mathematics (see Della Rocca 2016 for a compelling explanation). The philosophy of mathematics has seemed to have little to do with the philosophy of history, and the way in which history holds together the modalities of possibility, contingency, and necessity in human action. Even recent works that bring the history of mathematics to bear on the philosophy of mathematics in one way or another contain little philosophical discussion of history as such: they instead take their lead from current discussion in philosophy of science, where historical episodes are taken to be evidence for philosophical theses, or instantiations of them. That is, the relation between episodes in the history of mathematics to philosophical theses about mathematics is understood in terms of a scientific model of the relation between scientific theory and empirical fact. The problem with this model is that it is itself timeless and ahistorical, that is, it construes the history of mathematics as a set of facts that can support or instantiate (or falsify) a theory, and be explained (or predicted) by a theory; thus the very historicity of mathematics is lost to philosophical view (see Brown 2008; Gillies 1993; Kitcher 1983; Maddy 2000).

This difficulty is compounded by the epistemology that most philosophers of mathematics have chosen to work with to undergird their accounts of mathematics. The first, an empiricist 'naturalized epistemology', begins with an appeal to perceptual processes that, while temporal, are not historical. They are construed as processes in nature, like those described by physics or biochemistry: the impingement of light on organs, the transmission of electrical impulses in neurons, etc., so although they happen in time, they take place in the same way universally, in all times and locations. They are not historical because they do not share in the peculiarity, idiosyncrasy, and irrevocability of historical character. The other, rationalist alternative, truth as formal proof, evidently stands not only outside of history but of time as well.

Mathematics is famous for lifting human beings above the contingencies of time and history, into a transcendent, infinitary realm where nothing ever changes. And yet philosophy of mathematics requires philosophy of history, because the discovery of conditions of intelligibility takes place in history, and because we lend ourselves to the intelligibility of things in undertaking the search. If we can borrow Plato's simile of ascending from the cave and Leibniz's simile of attaining a wider and more comprehensive *point de vue* of a city without the totalizing supposition of The Good, or of God, we can begin to see how the search for conditions of intelligibility takes place in history. If we can borrow Hegel's dialectic or Peirce's creative evolution of the universe without the totalizing supposition of the end of history, then we can begin to see how we lend ourselves to the intelligibility of things. And if we can invoke reason or intelligibility without trying to set pre-established limits for it, then we might arrive at a philosophy of history useful for understanding mathematics.

What use should the philosophy of mathematics make of history? Philip Kitcher, in his book *The Nature of Mathematical Knowledge*, just cited, tries to bring the philosophy of mathematics into relation with the history of mathematics, as a quarter of a century earlier Kuhn, Toulmin, and Lakatos aimed to do for the philosophy of science. Yet his approach is quite ahistorical. The epistemology he offers has no historical dimension: it is, he claims, a defensible empiricism. "A very limited amount of our mathematical knowledge can be obtained by observations and manipulations of ordinary things. Upon this small basis we erect the powerful general theories of modern mathematics", he observes, hopefully, and adds, "My solution to the problem of accounting for the origins of mathematical knowledge is to regard our elementary mathematical knowledge as warranted by ordinary sense perception". He does admit that "a full account of what knowledge is and of what types of inferences should be counted as correct is not to be settled in advance..." especially since most current epistemology is still dominated by the case of perceptual knowledge and restricted to intra-theoretic reasoning (Kitcher 1983: 92–97).

However, his own epistemological preliminaries seem to be so dominated and restricted: "On a simple account of perception, the process would be viewed as a sequence of events, beginning with the scattering of light from the surface of the tree, continuing with the impact of light waves on my retina, and culminating in the formation of my belief that the tree is swaying slightly; one might hypothesize that none of my prior beliefs play a causal role in this sequence of events.... A process which warrants belief counts as a basic warrant if no prior beliefs are involved in it, that is, if no prior belief is causally efficacious in producing the resultant belief. Derivative warrants are those warrants for which prior beliefs are causally efficacious in producing the resultant belief". A warrant is taken to refer to processes that produce belief in the right way. Then "I know something if and only if I believe it and my belief was produced by a process which is a warrant for it" (Kitcher 1983: 18–19). This is an account of knowledge with no historical dimension. It also represents belief as something that is caused, for a basic warrant is a causal process that produces a physical state in us as the result of perceptual experience and which can (at least in the case of beliefs with a basic warrant) be engendered by a physical process.

In a later article, "Mathematical Progress", Kitcher makes a different, rather more pragmatic claim about mathematical knowledge, characterizing 'rational change' in mathematics as that which maximizes the attainment of two goals: "The first is to produce idealized stories with which scientific (and everyday) descriptions of the ordering operation that we bring to the world can be framed. The second is to achieve systematic understanding of the mathematics already introduced, by answering the questions that are generated by prior mathematics". He then proposes a concept of "strong progress", in which optimal work in mathematics tends towards an optimal state of mathematics: "We assume that certain fields of mathematics ultimately become stable, even though they may be embedded in ever broader contexts. Now define the limit practice by supposing it to contain all those expressions, statements, reasonings, and methodological claims that eventually become stably included and to contain an empty set of unanswered questions" (Kitcher 1988: 530–531). So there are two assumptions that render Kitcher's account ahistorical. One is that mathematical knowledge has its origins in physical processes that cause fundamental beliefs in us (and these processes, while temporal, are not historical). The other is that mathematics should optimally end in a unified, universal, axiomatized system where all problems are solved and have their place as theorems. This unified theory has left history behind, like Peirce's end of science or Hegel's end of history, so that history no longer matters.

The intervention of history between the ahistorical processes and objects of nature, and the ahistorical Ultimate System (related to the former as theory to model, thus by the ahistorical relationship of instantiation) seems accidental. Kitcher puts all the emphasis on generalization, rigorization, and systematization, processes that sweep mathematics towards the Ultimate System, with its empty set of unanswered questions. The philosophy of mathematics of Jean Cavaillès, provides an instructive contrast here. The method of Cavaillès, like that of his mentor Leon Brunschvicg, is historical. He rejects the logicism of Russell and Couturat, as well as the appeal of Brouwer and Poincaré to a specific mathematical intuition, referring the autonomy of mathematics to its internal history, "un devenir historique original" which can be reduced neither to logic or physics (Cavaillès 1937). In his historical researches (as for example into the genesis of set theory), Cavaillès is struck by the ability of an axiomatic system to integrate and unify, and by the enormous autodevelopment of mathematics attained by the increase of abstraction. The nature of mathematics and its progress are one and the same thing, he thinks, for the movement of mathematical knowledge reveals its essence; its essence is the movement.

Cavaillès always resists the temptation to totalize. History itself, he claims, while it shows us an almost organic unification, also saves us from the illusion that the great tree may be reduced to one of its branches. The irreducible dichotomy between geometry and arithmetic always remains, and the network of tranversal links engenders multiplicity as much as it leads towards unification. Moreover, the study of history reminds us that experience is work, activity, not the passive reception of a given (see Sinaceur 1994: 11–33; Sinaceur 2013: Chap. 6; see also Wagner 2016: Chap. 1). Thus for Cavaillès a mathematical result exists only as linked to both the context from which it issues, and the results it produces, a link which seems to be both a rupture and a continuity.

3 Reference in Mathematics

Current philosophical discussion of reference in mathematics is a bit hard to characterize. Sometimes the problem of model construction is substituted for the problem of reference; this move is favored by anti-realist philosophers. Thus theories are about models, though this leaves open the issue of how models refer, or if they do; and models are not individuated in the way that mathematical things typically are individuated. Bertrand Russell argued a century ago that the reference of a name is fixed by a proper definite description, an extensionally correct description which picks out precisely the person or natural kind intended (Russell 1905: 479–493). And W. V. O. Quine argued half a century ago that the ontological commitment of a first order theory is expressed as its universe of discourse (Quine 1953/1980: 1–19). But first order theories do not capture the objects they are about (numbers, figures, functions) categorically, and the 'ontological commitments' of higher order theories are unclear. Saul Kripke insisted that we need the notion of an initial baptism (given in causal terms), followed by an appropriate causal chain that links successive acts of reference to the initial act, for only in this case would the name be a 'rigid designator' across all possible worlds; a rigid designator picks out the correct person or natural kind not only in this world but in all possible worlds where the person or kind might occur (Kripke 1980). Hilary Putnam argued that the ability to correctly identify people and natural kinds across possible worlds is not possessed by individuals but rather by a society where epistemic roles are shared (Putnam 1975). And Paul Benacerraf argued in a famous essay that linking reference to causal chains makes an explanation of how mathematics refers seem futile (Benacerraf 1965).

In sum, it is not generally true that what we know about a mathematical domain can be adequately expressed by an axiomatized theory in a formal language; and it is not generally true that the objects of a mathematical domain can be mustered in a philosophical courtyard, assigned labels, and treated as a universe of discourse. What troubles me most about this rather logicist picture is that the difficulty of integrating or reconciling the two tasks of analysis and reference (as well as the epistemological interest of such integration) is not apparent, since it is covered over by the common logical notions of instantiation and satisfaction.

The assumption seems to be that all we need to do is assign objects and sets of objects from the universe of discourse (available as a nonempty set, like the natural numbers) to expressions of the theory. If we carry out the assignment carefully and correctly, the truth or falsity of propositions of the theory, vis-à-vis a 'structure' defined in terms of a certain universe of discourse, will be clear. In a standard logic textbook, the universe of discourse is the set of individuals invoked by the general statements in a discourse; they are simply available. And predicates and relations are treated as if they were ordered sets of such individuals. In real mathematics, however, the discovery, identification, classification and epistemic stability of objects are problematic; objects themselves are enigmatic. It takes hard work to establish certain items (and not others) as canonical, and to exhibit their importance. Thus reference is not straightforward. Moreover, of course, neither is analysis; the search for useful predicates and relations proceeds in terms of intension, not extension, and the search for useful methods and procedures cannot be construed extensionally at all. Analysis is both the search for conditions of intelligibility of things and for conditions of solvability of the problems in which they figure. We investigate things and problems in mathematics because we understand some of the issues they raise but not others; they exist at the boundary of the known and unknown.

My claim that mathematical objects are problematic (and so in a sense historical and in another sense strongly related to practices) need not lead to skepticism or anti-realism. We can argue, with Scott Soames *inter alia*, that natural language (English, French, German, etc.) provides us with an enormous amount of reliable information about the way the world is, what things are in it and how they turn up; and note that we act on this information as we reason (Soames 2007). Natural language, along with abstract mathematical structures, also allow us to bring the disparate idioms of mathematics (which are so different from natural language) into rational relation. So presumably mathematics enhances our information about the world, modifying without dismissing our everyday understanding. We can claim that discourse represents things well without becoming dogmatic, if we leave behind the over-simplified picture of the matching up of reference and analysis as the satisfaction of propositions in a theory by a structure.

Things, even the formal things of geometry and arithmetic, have the annoying habit of turning out to be irreducible to discourse. Things transcend discourse even though of course they lend themselves to it. But their irreducibility is what makes truth possible: we can't just make things up or say whatever we want about things. A true discourse must have something that it is about. The history of both mathematics and science shows that things typically exhibit the limitations of any given discourse sooner or later, and call upon us to come up with further discourses. Because of their stubborn *haeccity*, things are, we might say, inexhaustible. Thus when we place or discover them in a novel context, bringing them into relation with new things, methods, modalities of application, theories, and so forth, they require revised or expanded discourses which reveal new truths about them.

The irreducibility of things should also lead us to pay closer attention to what I have called subject-discourses, how they are constituted and how they function (given that they must point beyond themselves). Given their function, they must be truer to the individuality or specificity of things than predicate-discourses need to be, since the function of predicate-discourses is to generalize. So they will be less systematic and well-organized, but more expressive, messy, precise, and surprising.

Another way of putting this is that any discourse that encompasses a broad domain of problems in mathematics or science will be internally bifurcated, since it must integrate a subject-discourse and a predicate-discourse. Due to the inherent disparity of such discourses, this unification will become unstable sooner or later. Moreover, any discourse fails to exhaust knowledge about the things it concerns; sooner or later those problematic, resistant things, and the novel (worldly and discursive) contexts that may arise with respect to them, will generate problems that the original discourse can't formulate or solve. Indeed, the very desire to solve problems via problem-reduction may bring about the situations that force the revision, replacement and extension of discourse. We often place things in novel contexts, or come to recognize that novel contexts can be made to accommodate familiar things, in order to elicit new information. In sum, the model of theory reduction, examined in detail in Chap. 4, does not do justice to the internal bifurcation of both the reduced and reducing theories, and it does not capture the complex relations between them; and so it does not attend properly to the ways in which knowledge grows. Processes of problem-reduction should be examined by philosophers alongside processes of theory-reduction; and the ways in which these processes are ampliative should be recognized.

4 Wiles' Proof of Fermat's Last Theorem

Here I return to the story of Andrew Wiles and his proof of Fermat's Last Theorem. On the one hand, we have a narrative about an episode in the life of one man (in a community of scholars) who, inspired by a childhood dream of solving Fermat's Last Theorem, and fortified by enormous mathematical talent, a stubborn will, and the best number theoretical education the world could offer, overcame various obstacles to achieve truly heroic success. Indeed, the most daunting and surprising obstacle arose close to the end, as he strove to close a gap discovered in the first draft of his proof. On the other hand, we have a proof search which can be mapped out and reversed into the full dress proof, though it is important to recall that the proof is not located in any single axiomatized system. It makes use not only of the facts of arithmetic and theorems of number theory (both analytic number theory and algebraic number theory), but also results of group theory (specifically Galois theory); and it exploits the system of p-adic numbers (offspring of both topology and group theory), representation theory, deformation theory, complex analysis, various algebras, topology, and geometry. This proof search has its own location in history, which must be distinguished from that of Wiles' life, for it constitutes a path backwards through mathematical history (where earlier results make later results possible, and where new results bring earlier results into new alignments) and a leap that is also a rupture opening onto the future, making use of older techniques in novel ways to investigate a conjecture that many number theorists in fact worried could not be proved by the means available at the time.

Wiles' proof of Fermat's Last Theorem relies on verifying a conjecture born in the 1950s, the Taniyama-Shimura conjecture, which proposes that every rational elliptic curve can be correlated in a precise way with a modular form. (It is a nice example for Carlo Cellucci's philosophical approach, discussed in Chap. 3, because Fermat's Conjecture is true, if the Taniyama-Shimura conjecture is true, and this turns out to be a highly ampliative reduction.) It exploits a series of mathematical techniques developed in the previous decade, some of which were invented by Wiles himself. Fermat wrote that his proof would not fit into the margin of his copy of Diophantus' *Arithmetica*; Wiles' 108 pages of dense mathematics certainly fulfills this criterion. Here is the opening of Wiles' proof: "An elliptic curve over $\mathbf{Q}$ is said to be modular if it has a finite covering by a modular curve of the form $X_0(N)$. Any such elliptic curve has the property that its Hasse-Weil zeta function has an analytic continuation and satisfies a functional equation of the standard type. If an elliptic curve over $\mathbf{Q}$ with a given j-invariant is modular then it is easy to see that all elliptic curves with the same j-invariant are modular… A well-known conjecture which grew out of the work of Shimura and Taniyama in the 1950s and 1960s asserts that every elliptic curve over $\mathbf{Q}$ is modular… In 1985 Frey made the remarkable observation that this conjecture should imply Fermat's Last Theorem. (Frey 1986). The precise mechanism relating the two was formulated by Serre as the ε-conjecture and this was then proved by Ribet in the summer of 1986. Ribet's result only requires one to prove the conjecture for semistable elliptic curves in order to deduce Fermat's Last Theorem" (Wiles 1995: 443). For my brief exposition of the proof here, in Chap. 5 and in Appendix B, I relied upon the original article as well as didactic expositions and my own class notes (Darmon et al. 1997; Li 2001, 2012, 2013, 2014; Ribet 1995).

In number theory, as we have seen, the introduction of algebraic notation in the early seventeenth century precipitates the study of polynomials, algebraic equations, and infinite sums and series, and so too procedures for discovering roots and for determining relations among roots or between roots and coefficients, and ways of calculating various invariants. The use of abstract algebra (groups, rings, fields, etc.) in the late nineteenth and early 20th centuries leads to the habit of studying the symmetries of algebraic systems as well as those of geometrical items, finitary figures and infinitary spaces. The habit of forming quotients or 'modding out' one substructure with respect to its parent structure often produces a finitary structure with elements that are equivalence classes, from two quite infinitary structures. This habit in turn suggests the use of two-dimensional diagrams characteristic of (for example) deformation theory, where the relations among the infinitary (or very high-dimensional) and the finitary are displayed in what might be called iconic fashion. Abstract algebra also produces the habit of seeking in the relation of structure to substructure other, analogous relations in different kinds of structure and substructure. For example, the Fundamental Theorem of Galois Theory tells us that, when G is the Galois group for the root field N of a separable polynomial $f(x)$ over a field F, then there is a one-one correspondence between the subgroups of G and the subfields of N that contain F. And Representation Theory instructs us to seek groups of matrices that will mimic in important ways the features of other infinitary and less well-understood groups of automorphisms. Abstract algebra also suggests the investigation of a given polynomial over various fields, just to see what happens, as modern logic (treated algebraically) suggests the investigation of non-standard models, just to see what they are like.

So what is an aspect of reference for one number theorist, like Barry Mazur who takes his orientation from algebraic topology and cohomology theory, may have played the role of analysis for other, more traditional number theorists like Eichler and Shimura, who begin from the arithmetic theory of Abelian varieties. What preoccupies one number theorist may remain tacit for another, and vice versa, so that the combination of their results (as in the case of Wiles' proof) forces the articulation of ideas which had up till then remained out of focus, beyond the horizon of attention. Likewise, what remains tacit for the number theorist may be articulated by the logician, as we shall see in Chap. 5. For what remains tacit in one approach (given the strengths and limitations of a given mathematical idiom) must often be made explicit in another in order to bring the two approaches into productive relation, as novel strategies of integration are devised.

I will sketch the proof of Fermat's Last Theorem in terms of two stages, briefly. The first stage concerns the result of Eichler-Shimura, which proves that given a certain kind of modular form, we can always find a corresponding elliptic curve. (This stage is explained at length in Chap. 5, with a glossary (and adumbrated in Appendix B), and its philosophical implications explored.) The second stage concerns Wiles' result, proving the Taniyama-Shimura conjecture, that given a certain kind of elliptic curve, we can always find a certain kind of modular form. (To explain this stage, I would have to write another book; Appendix A offers some useful historical background.) Frey conjectured and Ribet proved that Fermat's Last Theorem follows from this correspondence, carefully qualified. (Ribet shows that the correspondence rules out the possibility of counterexamples to Fermat's Last Theorem; see Ribet 1990). The strategy that figures centrally in the Eichler-Shimura proof is the strategic use of L-functions (generalizations of the Riemann zeta function, and Dirichlet series), where given a certain kind of modular form *f* we have to construct a corresponding, suitably qualified, elliptic curve *E*. Another equally important strategy is to use representation theory in tandem with deformation theory, where *p*-adic families of Galois representations figure centrally in the proof of the Taniyama-Shimura conjecture. Given a certain kind of elliptic curve *E*, we investigate *p*-adic representations in order to construct a corresponding, suitably qualified, modular form *f*.

5 Wiles' Analysis Considered as History

Andrew Wiles' fascination with Fermat's Last Theorem began when he was 10 years old, and culminated on the morning of September 19, 1994, when he finally put the last piece of the grand puzzle in place. In order to establish the isomorphism between T_Σ and R_Σ, he had tried to use an approach involving 'Iwasawa theory', but that had been unsuccessful; then he tried an extension of the 'Kolyvagin-Flach method', but that attempt had stalled. While trying to explain to himself why this new approach didn't seem to be working, he realized (inspired by a result of Barry Mazer's) that he could use Iwasawa theory to fix just that part of the proof where the Kolyvagin-Flach approach failed; and then the problem would be solved. On that morning, something happened that was radically unforeseeable (even by Wiles, who was very discouraged and did not believe it would happen), and yet, once it actually took place, presented the kind of necessity that mathematical results present. It disrupted mathematics by changing its transversal relations, for now modular forms were proved to be correlated in a thoroughgoing way with elliptical curves, and at the same time established a new order. The unforeseeability was not merely psychological, subjective, and merely human; the disruption lay in the mathematical objects as well as in the mind of the mathematician.

What Wiles did on that morning can only be explained in terms of the mathematics. As Cavaillès argues, "I want to say that each mathematical procedure is defined in relation to an anterior mathematical situation upon which it partially depends, with respect to which it also maintains a certain independence, such that the result of the act [*geste*] can only be assessed upon its completion" (Cavaillès and Lautmann 1946: 9). A mathematical act like Wiles' is related to both the situation from which it issues and the situation it produces, extending and modifying the pre-existing one. It is both a rupture and a continuation, an innovation and a reasoning. To invent a new method, to establish a new correlation, even to extend old methods in novel ways, is to go beyond the boundaries of previous applications; and at the same time in a proof the sufficient conditions for the solution of the problem are revealed. What Cavaillès calls the fundamental dialectic of mathematics is an alliance between the necessary and the unforeseeable: the unforeseeability of the mathematical result is not appearance or accident, but essential and originary; and the connections it uncovers are not therefore contingent, but truly necessary.

Another way of describing Cavaillès' insight is to say that he is trying to uncouple the connection between necessity and the Kantian a priori, which offers only Kantian analysis, the unpacking of what is already contained in a concept, or synthesis, which must be referred to the mind of the knower. What happened when Wiles finally proved Fermat's Last Theorem? Just at that point, the unsolved problem was solved, the unforeseeable flipped over and was seen at last, the indeterminately possible became the determinately necessary. It was at once an event in the biography of Andrew Wiles: his alone was the consciousness in which this amazing peripety or reversal took place, this discovery, a change from ignorance to knowledge. No one else could have shared that discovery as it happened for the first time, that singular event, for no one can inhabit the mind of another: as Leibniz said, we are windowless monads. Yet both the dramatic structure the act already possessed, and the argumentative structure inherent in the proof underlay the story Wiles recounted over and over the next day, to himself (checking the proof), then to his wife, then to his colleagues, then to the world, in different fashions. And the story-argument didn't change thereafter, as it was reenacted in the thoughts of those mathematicians who knew enough number theory to check the proof, all of whom found it successful.

The retelling, which includes his narrative of the proof-search, and the published 108 page proof, is marked by the idiosyncracy and irrevocability of that historical moment in one obvious way: *Wiles could only make use of results that had been discovered up to that point in history when he finished devising his proof.* The Taniyama-Shimura conjecture requires the availability of modular forms, which rests on the work of Poincaré in exhibiting their infinite symmetry, which requires the work of Klein and Riemann in formulating hyperbolic space as devised by Lobachevsky. What is requisite for formulating and solving a problem lies only in the past, made available by instruction or textbook. Fermat could formulate the problem of the Last Theorem, but despite his boast he could not have solved it.

6 The Math Genie

The incoherence of the notion of an Ideal Chronicle bears not only on the reality of human beings and their acts, but on the possibility of giving a complete speech about the totality of human action because the very description of an event is interpretive, because one cannot eliminate from the description of events terms that link it to both its past and its future, and because the description of an event changes with time as the event comes to stand in different relations to events that come after it. Likewise, the notion, suggested by Kitcher, of some Ultimate System in mathematics is also incoherent, not because the reality or intelligible unity of mathematical things is doubtful, but because we cannot give a complete speech or formal theory about them. The very description of a mathematical object is interpretive, because it is given against a background of antecedent knowledge and by means of a certain notation; one cannot eliminate from its description terms that link it to past problems and problems still to be solved; and the relations and correlations in which it stands to other objects change over time, as new objects are discovered and older objects are forgotten.

Like perceived things, mathematical things are problematic. Just as perceived things call for analysis in order to uncover the conditions of their intelligibility, so mathematical things call for analysis in order to uncover the conditions of solvability of the problems in which they are always embedded. But to be problematic is an historical feature. Objects are problematic when we understand enough about them to see with some precision what we don't yet know about them. And as soon as we learn something new about them, in virtue of that very discovery they typically come to stand in novel, unforeseen relations with other objects that make them problematic again. As I showed in Chap. 1, when the classical problems concerning the circle were solved during the 17th century by the new analytic geometry and the infinitesimal calculus, those same discoveries relocated the circle in relation to transcendental curves (especially sine and cosine), the definition of curvature and the generalized notion of a surface, etc., and re-embedded it in a host of new problems. We know what we know about the circle up to this point in history, with the means at our disposal as those means have been deployed; and we can't yet know other things about it, though we can question, postulate, conjecture, hypothesize, acts that project us towards that future though we are not quite yet there. Asking questions and making conjectures is a way of approaching knowledge that we do not yet have: so analysis is not just a pathway into the history of mathematics (though it is that), but also an unfinished bridge to the future. The relation between a problem and its conditions of solvability may be lifted, as it were, out of history: this is what we do when we turn the search for a proof into a proof. Yet the fact that the problem was a problem and is now a solved problem, is a fact that belongs as much to history as to mathematics. Insofar as every theorem may be said to be a solved problem, the same holds true for theorems (see Hersh 1999).

There are no problems without problematic things; problems exist in mathematics because we encounter things that trouble us. What would a problem be about if it weren't about some thing? What would a problem be without its aboutness? Versions of Fermat's Last Theorem existed before Fermat, as a range of problems about positive whole numbers (or rather, triples of them); Fermat turned it into a problem about a set of polynomial equations; Andrew Wiles turned it into a problem about a correlation between elliptic curves and modular forms. The evolution of the problem depends on rational relations among different kinds of objects: numbers serve as conditions of intelligibility for polynomial equations, and the latter for the Taniyama-Shimura correlation. Wiles' result can be read backwards, to hold for the equations and the relevant number-triples; the aboutness of a problem may change, but the aboutness-apropos-earlier-things remains as a condition of the intelligibility of the problem. And there are no things without problems. Things, even they serve as conditions of intelligibility of other things, don't wear the conditions of their own intelligibility on their faces; it is always a problem for reflection to find the conditions of intelligibility of a thing.

Indeed, if a mathematician could magically reach into the future and bring back future results, then there would be no activity called solving problems. The appearance of mathematical problems *as* problems requires history, and history demands our patience as we wait to see how things will turn out. Let us suppose that there is an Ultimate System, a complete system of mathematics independent of the accidents of history, presided over by the Math Genie. This genie comes to the aid of mathematicians and maybe even philosophers who rub the relevant lamp: he brings the solution to any problem that troubles you in the form of information about the objects involved. He can violate history because he has access to the Ultimate System. Imagine what this genie might have done for Fermat in the mid-seventeenth century, when in fact the mathematical resources for proving his Last Theorem were lacking! He would have set out Andrew Wile' proof for him, but of course to ensure that Fermat understood the proof, the genie would have had to teach him, perhaps in a series of seminars, all the 18th, 19th, and 20th c. mathematics linking what he knew to what Wiles knew, perhaps keeping him alive by philtres until the process could be completed. In this case, the genie would have to offer information, not just about numbers and equations, but about the correlation between modular forms and elliptic curves. He would have to bring numbers into rational relation with objects that, given the constraints of 17th c. mathematics, were not even thinkable.

However, we have just told the story from the point of view of the Math Genie (and tacitly assumed that he is located in our era, cleverly disguised as eternity). We must tell it from Fermat's point of view; but then we see that Fermat could not have made his request successfully in the first place. Suppose that Fermat had asked the Math Genie to bring back the solutions to his problem, as he himself enunciated it, for the genie might have required that all requests be precisely specified. In that case, he could hardly have brought back Wiles' result, for though Wiles showed that it entailed a proof of Fermat's Last Theorem, Fermat could never have asked the genie for a proof of the Taniyama-Shimura conjecture and its reduction to his problem. Alternatively, the genie might have acceded to his general request to bring back problems related to the natural numbers: but then he would have had to go into the future and bring back all such problems involving all the new objects that include the natural numbers in their genealogy. This is a limitless prospect: by now we have seen enough of analysis to know that an analysis typically uncovers new objects and problems: there is always tacit knowledge at the metalevel to enunciate, new generalizing abstractions to create (and thereby to lose or forget other things), new correlations to explore, and so forth. Fermat would have been swamped.

So either Fermat can't ask for what he wants; or to the extent that he can ask for it, the genie can only offer him an unsurveyable infinity, without any kind of closure, of solved problems. The incoherence of supposing that mathematics can be liberated from the accidents of history, that all problems might be solved and an Ultimate System projected in which "the set of unanswered problems would be empty" shows that the historical location of a problem and the way in which the objects involved in it are problematic is not accidental but essential to the problem as such. Problems can only appear in a situation where some things are known and some things are not yet known; the enunciation of a problem is just saying what precisely is not known against the background of what has been discovered so far, and suggestions about how to proceed to solve the problem require even greater precision. Moreover, problems and their solutions are the articulation of mathematics: they provide it with the intelligible structure that may be written afterwards as theorems and axioms that organize theorems. The Math Genie is a useful fiction, like Descartes' Evil Demon, to show the incoherence of an idea, in this case, that of mathematics without problems. The thought experiment just entertained shows at least two reasons why there cannot be a complete speech about mathematical things, any more than there can be a complete speech about human actions. (And this is no more a reproach to the reality of mathematical things than it is to the reality of who we human beings are and what we do.) One is that there are many different kinds of mathematical things, which give rise to different kinds of problems, methods, and systems. The other is that mathematical things are investigated by analysis, which is a process at once logical and historical in which some things, or features of things, that were not yet foreseen are discovered, and others are forgotten. Indeed, these two aspects of mathematics are closely related. For when we solve problems, we often do so by relating mathematical things to other things that are different from them, and yet structurally related in certain ways, as when we generalize to arrive at a method, or exploit new correlations. We make use of the internal articulation or differentiation of mathematics to investigate the intelligible unities of mathematics. To put it another way, just as there is a certain discontinuity between the conditions of solvability of a problem in mathematics and its solution (as Cavaillès noted), so there is a discontinuity between a thing and its conditions of intelligibility (as Plato noted). An analysis results in a speech that both expresses, and fails to be the final word about, the thing it considers.

References

Aristotle. (1947). *Introduction to Aristotle*. R. McKeon (Ed.). New York: Random House.

Benacerraf, P. (1965). What numbers could not be. *Philosophical review, 74*(1), 47–73.

Breger, H. (2000). Tacit knowledge and mathematical progress. In E. Grosholz & H. Breger (Eds.), *The growth of mathematical knowledge* (pp. 221–230). Dordrecht: Kluwer.

Brown, J. (2008). *Philosophy of mathematics*. Oxford: Routledge.

Cavaillès, J. (1937). Réflections sur le fondement des mathématiques. In *Travaux du IVe Congrès international de philosophie*, VI/535. Paris: Hermann.

Cavaillès, J., & Lautmann, A. (1946). La pensée mathématique. *Bulletin de la Société française de philosophie, 40*(1), 1–39.

Cellucci, C. (2013). *Rethinking logic: Logic in relation to mathematics, evolution and method*. Dordrecht: Springer.

Danto, A. (1965). *The analytical philosophy of history*. Cambridge: Cambridge University Press.

Darmon, H., Diamond, F., & Taylor, R. (1997). Fermat's last theorem. In *Conference on Elliptic Curves and Modular Forms* (pp. 1–140), Dec. 18–21, 1993. Hong Kong: International Press, 1997.

Della Rocca, M. (2016). Meaning, the history of philosophy, and analytical philosophy: A parmenidean ascent. In M. van Ackeren (Ed.), *Philosophy and the historical perspective. Proceedings of the British Academy*. Oxford: Oxford University Press.

Frey, G. (1986). Links between stable elliptic curves and certain diophantine equations. *Annales Universitatis Saraviensis, 1*, 1–40.
Gallie, W. B. (1964). *Philosophy and the Historical Understanding*. New York: Schocken.
Gardiner, P. (1961). *The Nature of Historical Explanation*. Oxford: Oxford University Press.
Gillies, D. (1993). *Philosophy of science in the twentieth century*. Oxford: Blackwell.
Hersh, R. (1999). *What is mathematics, really?*. Oxford and New York: Oxford University Press.
Ippoliti, E. (2008). *Inferenze ampliative*. London: Lulu.
Kitcher, P. (1983). *The Nature of Mathematical Knowledge*. New York: Oxford University Press.
Kitcher, P. (1988). Mathematical progress. *Revue Internationale de Philosophie, 42/167*(4), 518–531.
Kripke, S. (1980). *Naming and Necessity*. Princeton: Princeton University Press.
Li, W-C. (2001, 2012, 2013, 2014). Class notes, recorded by Emily Grosholz, Pennsylvania State University.
Lucretius. (2007). *The nature of things* (A. E. Stallings, Trans.). London: Penguin.
Maddy, P. (2000). *Naturalism in mathematics*. New York: Oxford University Press.
Plato. (1961). *The collected dialogues of Plato including the letters*. New York: Bollingen Foundation.
Putnam, H. (1975). *Mind, language and reality*. New York: Cambridge University Press.
Quine, W. V. O. (1953/1980). *From a logical point of view*. New York: Harper.
Ribet, K. (1990). On modular representations of Gal (Q/Z) arising from modular forms. *Inventiones Mathematicae, 100*, 431–476.
Ribet, K. (1995). Galois representations and modular forms. *Bulletin of the American Mathematical Society, 32*(4), 375–402.
Russell, B. (1905). On denoting. In *Mind, 14*(56).
Sinaceur, H. (1994). *Jean Cavaillès, Philosophie mathématique*. Paris: PUF.
Sinaceur, H. (2013). *Cavaillès*. Paris: Les Belles Lettres.
Soames, S. (2007). *Reference and description: The case against two-dimensionalism*. Princeton: Princeton University Press.
Wagner, R. (2016). *Making and breaking mathematical sense: Histories and philosophies of mathematical practice*. Princeton: Princeton University Press.
Wiles, A. (1995). Modular elliptic curves and Fermat's last theorem. *Annals of Mathematics, 141* (3), 443–667.

Chapter 3
Rethinking Ampliative Reasoning

What can a philosopher of mathematics say about ampliative reasoning, in order both to acknowledge the importance of the canons of deductive logic, and to explain why reason always seems to go beyond those canons? In this chapter, I briefly explore two historical episodes, the investigation of a certain transcendental curve (the catenary), and the early development of complex analysis, which highlight the disparity of reference and analysis in mathematical research, and how difficult and rewarding it is to address that disparity; and in each section I bring in the interesting ideas of a few of my contemporaries (with whom I agree and disagree) to develop the exposition: philosophers of mathematics Karine Chemla, Carlo Cellucci, and Danielle Macbeth, and philosophers of education Kieran and David Egan. In the following two chapters, I examine a series of case studies drawn from number theory to further develop and defend my ideas about analysis and ampliative reasoning in mathematics. I chose to work on number theory because even today many philosophers still suppose that the great vision of theory reduction that we find in the works of Russell and Whitehead, and Hempel and Carnap, is at least plausible as far as arithmetic is concerned. I will show that the constricting vision of arithmetic as a set of truths deducible from a set of axioms not only cannot account for the development of number theory, it also cannot even account for schoolroom arithmetic in the sense intended, even though logic sheds interesting light on arithmetic and number theory more generally. But the interplay between logic, arithmetic and number theory makes much more sense, and accords better with the practice of mathematicians, when it is understood as problem-solving that increases the content of mathematical knowledge, in logic, set theory, and number theory, in part because of the disparity among those highly specialized languages.

Moreover, the interaction of number theory with real and complex analysis, algebra, and topology continues blithely and productively today without concern for the strictures of formal logic, and yet does not go astray. Methods of problem solution are local (dependent on the peculiar features of certain kinds of mathematical objects, modalities of application, iconic conventions and symbolic idioms) and hybrid (crossing fields and exploiting polyvalences and ambiguities, a crossover

E.R. Grosholz, *Starry Reckoning: Reference and Analysis in Mathematics and Cosmology*, Studies in Applied Philosophy, Epistemology and Rational Ethics 30, DOI 10.1007/978-3-319-46690-3_3

that typically requires certain kinds of abstraction) and generalizing (invoking new, larger domains in which the problem-solving can proceed). So there is a dialectic between the local and the general, the referent and its analytical surround, upstairs and downstairs, the original idiom and the translation, in processes of discovery in mathematics, which ironically combines (as Cavaillès saw) with the discovery of necessary truths. But the presence of necessary truths stems from the peculiarly determinate and formal nature of mathematical things, and does not mean that deductive logic (with its own specific kinds of necessity) is the correct and single vehicle for the expression of mathematical reasoning. It only means that deductive logic is a branch of mathematics, like set theory and category theory; we have to give up the idea that those branches can provide mathematics with foundations or that mathematics needs foundations. If we pay attention to the problem of reference, we see that the various idioms of mathematics provide powerful but intensely local tools (in one sense) and models (in another sense) for solving problems. They only work well for representing certain kinds of things and helping to solve certain kinds of problems; to be transported, they must also be transformed, and the language by which that transformation may be conveyed is natural language. Oddly, natural language is very limited as a means of expression for mathematical problems; but we need to use it to bring disparate mathematical idioms into effective partnerships for solving problems. Thus, in principle there cannot be a single idiom for the expression of mathematics, much less mathematical physics.

We can see this in the pages of Descartes' *Geometry*, where French expository prose (and later, van Schooten's Latin) brings the import of the diagrams, ratios and proportions, formalized tracing instruments, and polynomial equations into rational relation. Problem-solving often occurs when a problem arises in one domain, but cannot be solved there, so that other domains are annexed in the service of problem-solving; the unification of arithmetic and geometry by Descartes' new algebra is a good example. Numbers (natural numbers, integers, rationals and algebraic numbers all figure in this narrative at different stages) are different in kind from geometric figures. To solve problems related to geometric figures, the Greeks reasoned in diagrams, establishing ratios and proportions among figures and parts of figures. To solve problems related to arithmetic, early modern Europeans learned to calculate using Indo-Arabic numerals and the techniques we all learn in primary school, which depend on the positional notation that allows a mere ten symbols (including 0) to stand for all the natural numbers. Descartes introduced a novel way of correlating numbers and line segments; his demonstrations exploit this novelty as well as Euclidean results. Thus not only do they juxtapose geometric figures with ratios and proportions, and polynomials (new objects precipitated by algebra) and equations; they superimpose interpretations onto the diagrams so that they must be read at once as about relations among line segments but also as about the triangles and circles of Euclid. The line segments are read at once as geometrical (Euclidean) items, as numbers (this transforms the concept of number), and as placeholders (this motivates the acceptance of polynomials as objects in their own right) (see Grosholz 1991: Chaps. 1 and 2). Thus polynomials, later on in number theory, come to function as very important links between different fields: over the rationals, a polynomial equation may mean a set of discrete points, while over the real Euclidean plane it means a curve and over the complex plane it means a Riemann surface.

1 Chemla: The Catenary

In her edition and translation (into French) of *Les neuf chapitres*, the Chinese mathematical classic, Karine Chemla argues that the organization of the book should not be read as a mere compendium of problems which exhibits a regretable 'lack of axiomatization' (see Chemla and Shuchun 2004; Grosholz 2005). Its strikingly non-Euclidean organization on the contrary deserves careful study, for it is rather the careful exploration of the meaning of procedures and algorithms in terms of problems that exhibit their correctness, clarify their domain of application, and indicate how that domain might be extended. Moreover, the emphasis is on explaining why a result is correct, not just demonstrating that it is correct. The working out of problems thus typically exhibits intermediary steps in a process of reasoning that contributes to the meaning of the final result, and indicates how one might go on from there. The study of Chemla's ideas about the Chinese tradition a decade ago helped me to see Leibniz's notion of analysis as an art of both discovery and justification in mathematics, which aims for generalization rather than abstraction, and explanation rather than formal proof. This characterization may seem odd to those whose view of Leibniz as the champion of formal proof was shaped by Bertrand Russell and Louis Couturat, and the later 20th c. philosophers who interpreted Leibniz under their influence. However, an attentive reading of Leibniz's own practice as a mathematician supports my claim, as do many of his philosophical reflections on that practice. I have made this case in an earlier book, apropos his treatment of various transcendental curves (Grosholz 2007: Chap. 8). And to this exploration I return here.

When Leibniz investigates a novel transcendental curve, he treats it as what Chemla would call a canonical object, in order to exhibit procedures or the algorithms that can be elicited from them. The exhibition of the meaning of procedures and the correctness of algorithms in terms of paradigmatic problems and canonical objects for Leibniz typically involves the combination of distinct modes of representation, including figures that exhibit spatial articulation, and descriptions of how to reason both upwards and downwards. It is a progressive and pedagogical search for the reasons that underlie general procedures and the constitution of objects, a search for deeper as well as broader understanding.

For Leibniz, the key to a curve's intelligibility is its hybrid nature, the way it allows us to explore numerical patterns and natural forms as well as geometrical patterns; he was as keen a student of Wallis, Huygens, and Cavalieri as he was of Descartes. These patterns are variously explored by counting and by calculation, by observation and tracing, and by applications of the language of ratios and proportions on the one hand and the new algebra on the other. To think them all together in the way that interests Leibniz requires the infinitesimal calculus as an *ars inveniendi*. The excellence of a characteristic for Leibniz lies in its ability to reveal conditions of intelligibility: for a transcendental curve, those conditions are arithmetical, geometrical, mechanical, and algebraic. What Leibniz discovers is that this 'thinking-together' of number patterns, natural forms, and figures, where his powerful and original insights into analogies pertaining to curves considered as hybrids can emerge, rebounds upon the very algebra that allows the thinking-together and changes it. The addition of the new operators d and $\int$, the introduction of variables as exponents (which changes the meaning of the variables), the consideration of polynomials as meaningful objects in themselves, and the entertaining of polynomials with an infinite number of terms, are all examples of this. Indeed, the names of certain canonical transcendental curves (log, sin, sinh, etc.) become part of the standard vocabulary of algebra and analysis.

This habit of generalization is evident throughout Volume I of the VII series (*Mathematische Schriften*) of Leibniz's works in the Berlin Akademie-Verlag edition, devoted to the period 1672–1676 (Leibniz 1990). As M. Parmentier admirably displays in his translation and edition *Naissance du calcul différentiel, 26 articles des* Acta eruditorum, the papers in the *Acta Eruditorum* taken together constitute a record of Leibniz's discovery and presentation of the infinitesimal calculus. These papers should be read in the context of Parmentier's French translation of Eberhard Knobloch's important edition of *De quadratura arithmetica circuli ellipseos et hyperbolae cujus corollarium est trigonometria sine tabulis*, where the integration of certain conics are shown to generate transcendental curves (see Leibniz 1989, 1993, 2004; Debuiche 2013a, b). These texts should be read not just as the exposition of a new method, but as the investigation of a family of related canonical items, that is, algebraic curves as well as transcendental curves, which prove to be of supreme importance for the interaction of mathematics and physics in the next two centuries. A new taxonomy of curves arises in these pages, where sequences of numbers alternate with geometrical diagrams accompanied by ratios and proportions, and with arrays of derivations carried out in Cartesian algebra augmented by new concepts and symbols.

For example, "De vera proportione circuli ad quadratrum circumscriptum in numeris rationalibus expressa," which treats the ancient problem of the squaring of the circle, moves through a consideration of the series $\pi/4 = 1 - 1/3 + 1/5 - 1/7 + 1/9\ldots$ to a number line designed to exhibit the finite limit of an infinite sum. Various features of infinite sums are set forth, and then the result is generalized from the case of the circle to that of the hyperbola, whose regularities are discussed in turn. The numerical meditation culminates in a diagram that illustrates the reduction: in a circle with an inscribed square, one vertex of the square is the point of intersection of two perpendicular asymptotes of one branch of a hyperbola whose point of inflection intersects the opposing vertex of the square. The diagram also illustrates the fact that the integral of the (algebraic) hyperbola is the (transcendental) logarithm. Integration takes us from the domain of algebraic functions to that of transcendental functions; this means both that the operation of integration extends its own domain of application (and so is more difficult to formalize than differentiation), and that it brings the algebraic and transcendental into rational relation (Leibniz 1682, 1989: 61–81).

Throughout the 1690s, Leibniz systematically investigates mathematics in relation to mechanics, deepening his command of the meaning and uses of differential equations, transcendental curves and infinite series. In the "Tentamen Anagogicum," Leibniz discusses his understanding of variational problems, fundamental to physics since, as he hypothesizes, all states of equilibrium and all motion occurring in nature are distinguished by certain minimal properties; his new calculus is designed to express such problems and the things they concern. The catenary is one such item; indeed, for Leibniz its most important property is the way it expresses an extremum, or, as Leibniz puts it in the "Tentamen Anagogicum," the way it exhibits a determination by final causes that exist as conditions of intelligibility for nature (Leibniz 1962, VII: 270–279). The catenary also turns out to be the evolute of the tractrix, another transcendental curve of great interest to Leibniz; thus it is intimately related to the hyperbola, the logarithmic and exponential functions, the hyperbolic cosine and sine functions, and the tractrix; and, of course, to the catenoid and so also to other minimal surfaces.

Leibnizian analysis lends itself to generalization (in Chemla's sense) in mathematics. Generalization starts from a set of solved problems, and asks how successful procedures may be extended to new problems and how their success may be explained. Problems involve problematic items, mathematical things that are intelligible but not wholly understood: a right triangle may be well defined, but we are still far from understanding the relation among its legs and hypotenuse. We may understand the relation among its legs and hypotenuse, but we are still far from understanding how a family of right triangles inscribed in a circle can define the transcendental functions sine and cosine. We may express the relation among its legs and hypotenuse in an equation, but we are still far from understanding the conditions under which whole number solutions to that equation may or may not exist. In the course of mathematical history, certain items (like the right triangle and the circle) prove to be canonical; canonicity is not an intrinsic quality that we discern by a sixth sense, by 'intuition,' but a feature of items that we discover in mathematical practice. Transcendental curves like sine, cosine and logarithm, and the catenary and tractrix, were not canonical for the Greeks, who were aware of just a few transcendental curves, as it were by accident. They became canonical only in the 17th century, once their important geometrical and mechanical properties were discerned, and their relations to each other and to the conic sections were studied, by means of novel mathematical idioms.

Leibniz characterizes his own mathematical practice vis-à-vis Descartes as a generalizing search for the conditions of intelligibility of canonical items, and the conditions of solvability of the problems in which those items are involved. In the introductory paragraph of his essay on the catenary, "De la chainette," he writes, "The ordinary analysis of Viète and Descartes consisted in the reduction of problems to equations and to curves of a certain degree… M. Descartes, in order to maintain the universality and sufficiency of his method, contrived for that purpose to exclude from geometry all the problems and all the curves that one couldn't treat by that method, under the pretext that they were only mechanics." This kind of exclusion, however, cuts off the process of generalization artificially, and Leibniz criticizes it. "But since these problems and curves can indeed be constructed or imagined by means of certain exact [tracing] motions, and since they have important properties and since nature makes frequent use of them, one might say that he [Descartes] committed an error in doing this, rather like that with which we reproach some of the Greeks, who restricted themselves to constructions by ruler and compass, as if all the rest were mechanics" (Leibniz 1692, my translation).

By contrast, Leibniz is inspired by the cogency and urgency of the excluded problems and curves to look for ways of expressing them in useful form and discovering conditions of solvability for them; he baptizes them 'transcendental' problems and curves, because they go beyond ordinary algebra. "This is what he [Leibniz] calls *the analysis of infinites*, which is entirely different from the geometry of indivisibles of Cavalieri, or Mr. Wallis' arithmetic of infinites. For the geometry of Cavalieri, which is by the way very restricted, is attached to figures, where it seeks the sums of ordinates; and Mr. Wallis, in order to facilitate research, gives us by induction the sums of certain sequences of numbers: by contrast, the new analysis of infinites doesn't focus on figures or numbers, but rather on general magnitudes, as ordinary algebra does. But it [the new analysis] reveals a new algorithm, that is, a new way to add, subtract, multiply, divide and extract roots, appropriate to incomparable quantities, that is, to those which are infinitely big or infinitely small in comparison with others. It employs equations involving finite as well as infinite quantities, and among those that are finite, allows equations with exponents that are unknowns, or rather, instead of powers and roots, it makes use of a novel appropriation of variable magnitudes, which is variation itself, indicated by certain characters, and which consists in differences, or in the differences of differences of certain degrees, to which the sums are reciprocal, as roots are to powers" (Leibniz 1692, my translation).

One can react to the transcendence or sublimity or inhumanity of mathematical things by turning against the situation. One can become a sophist, content to shift appearances at the first level of the Divided Line, or a materialist, content to remain with the sensible at the second level, or an empiricist, who hopes that the third level of the line is just an abstractive or constructive extension of the second. And these are natural choices, to try to shelter oneself within the ambit of the human, the computable, the visible. But philosophers who try to rest within what can be encompassed by finitary construction and perception in the end somehow never rest easy, and in any case never do justice to mathematics. They are plagued by difficulties (revealed by reductio arguments), unanswerable questions (revealed by burden of proof arguments), and by the way even the constructions and perceptions to which they cling seem willy-nilly to point beyond themselves. The other option, besides fleeing reason altogether, is to engage in analysis, that is, to recognize the top half of the Divided Line and try to find a philosophical speech about it, which acknowledges that it is real, that it stands in rational relation to us, and that its reality is different from ours. Even the finitary parts of mathematics involve the infinitary and even the visible world-system involves the invisible, as conditions of intelligibility.

The project of philosophical analysis is also self-reflective: it puts into question the relationship of the human mathematician (or philosopher) to the things of mathematics. We must not only try to explain the way in which knowledge unifies but also the way in which it holds things apart. Human beings, and their perceptions and constructions, are finitary and fleeting; but the things of mathematics are infinitary and eternal. If we try to reduce mathematics to merely human terms, to perceptions or finitary constructions, we avoid rather than confront the question of our rational relation to it, and the differences involved in that relation. In a sense, such confrontation—which sets us beside the working mathematician—is an exercise of our freedom, something about us which is after all infinitary and eternal (see Vuillemin 1997). If we pay attention to how differentiation and rational relatedness accompany each other, then not only the systematic unity of mathematics becomes salient but also the way in which systematic unities change over time as mathematicians explore the analogical likenesses of unlike things, or conversely try to articulate and conserve the intrinsic features of things apart from pressures exerted upon them by impinging analogies, or finally even try to consider likenesses in themselves, as novel methods and systems precipitate new items. In other words, history of mathematics becomes pertinent to philosophy of mathematics.

Logic, especially the expressive and subtle instrument of modern predicate logic, helps philosophers to examine mathematical systems considered as fixed and stable; but history provides evidence to help us understand how and why mathematics changes, and why it always moves beyond systematization. More generally, history is pertinent to the question of how mathematics brings very different kinds of things into determinate but revisable rational relation, and how this bears upon our philosophical understanding of rationality itself. Mathematics inspires us because it is at once inhuman and intelligible; it outstrips our finitary powers of construction, our perception, even our logic, at every turn, and nevertheless guides us because it stands in rational relation to us. The things of mathematics are problematic and yet intelligible, severely constraining what we can say about them; however, because they are so determinate, they render the little that we do manage to say about them necessary. Any finite thing in mathematics is at the same time an expression of the infinite, and the infinitary things that occur in our mathematics find finitary expression. This tension between the infinite and the finite in all of mathematics ensures that our knowledge, despite its precision, must remain incomplete. Mathematics also stands at the crossroads of history and logic: essential as logic is to the articulation of relations among mathematical items, the very constitution of a problem in mathematics is historical, since problems constitute the boundary between the known and the yet-to-be-discovered. We cannot explain the articulation of mathematical knowledge into problems and theorems without reference to both logic and history.

The philosophical reconstruction of mathematical practice is thus a delicate task, which requires scholarly familiarity with the detail of the mathematics of a given period, as well as the imaginative ability to go beyond the perspective of the mathematicians of that period and a respectful sense of their rationality. The problem with logic is its penchant for totalization and its intolerance of history; the outcome of mathematical progress is not always, and perhaps only rarely, an axiomatized system, where solved problems recast as theorems follow deductively from a set of special axioms, logical principles, and definitions. Careful study of the history of mathematics, even 20th c. mathematics, may discover that mathematicians pursue generality as often as they pursue abstraction, and sometimes prefer deeper understanding to formal proof. An axiomatic system is not the only model of theoretical unity, and deduction from first principles is not the only model for the location and justification of mathematical results.

2 Cellucci and Macbeth: Analysis and Productive Idioms

Carlo Cellucci began his career as a logician: he was Professor of Logic at the University of Rome La Sapienza for thirty years, and has only recently retired. However, very early on he began to chafe at the restrictions of mathematical logic as 20th c. philosophers applied and understood it, and to criticize the limited view of reasoning in mathematics and the sciences those applications engendered. His critique of the dogmas of logicism was accompanied by a search for alternative views, which he found in the history of logic and philosophy (the method of analysis) and in modern anthropology and neuroscience (a form of naturalism). In four books published in Italian over a period of ten years with the distinguished Roman press Laterza, he offers his alternative to the project of modern logic (*Le ragioni della logica*), his alternative to mid-twentieth century philosophy of mathematics (*Filosofia e matematica*), a detailed critique of most of the major schools of the latter (*La filosofia della matematica del Novecento*), and an account of his version of 'naturalized epistemology' (*Perché ancora la filosofia*). A presentation of the full range of his ideas was recently published in English, *Rethinking Logic: Logic in Relation to Mathematics, Evolution and Method* (Cellucci 1998, 2002, 2007, 2008, 2013).

The most important aspect of his work, I would argue, is his historical and philosophical account of the method of analysis. As Chemla works out her ideas in reference to classical Chinese mathematics, Cellucci re-reads the works of Plato and Aristotle, noting their radical conception of analysis: it is a method of problem-solving which is first and foremost ampliative. The method of analysis enlarges knowledge, going beyond what is currently accepted and what is given in the formulation of the problem. Thus he writes, "The analytic method is the method according to which, to solve a problem, one looks for some hypothesis that is a sufficient condition for solving it. The hypothesis is obtained from the problem, and possibly other data already available, by some non-deductive rule, and must be plausible, in a sense to be explained below. But the hypothesis is in its turn a problem that must be solved, and is solved in the same way… Thus solving a problem is a potentially infinite process" (Cellucci 2013: 5).

The heart of Cellucci's critique of 20th c. logicism is that it seeks to confine reason in the closed box of an axiomatic system within which we must reason deductively and where nothing new can ever be discovered; by contrast, the analytic method requires and expands an open space where research may and indeed must extend indefinitely. Problem-solving, he argues, not only in the empirical sciences but also in mathematics, is inherently ampliative. Moreover, axioms can never be secured by 'intuition,' and the great mid-century meta-theorems of mathematical logic, specifically the theorems of Gödel, place severe limits on its ability to provide foundations. Thus Cellucci rejects the aims and intent of mathematical logic, which he attributes to Frege (perhaps unfairly, see below), and summarizes as follows. The purpose of mathematical logic is to provide a secure foundation for mathematics, while correctly modeling the method of mathematics, that is, the axiomatic method. Thus philosophers and logicians need only be concerned with deductive justification, and can ignore the processes of discovery as merely psychological, irrational and idiosyncratic, though in a pinch one may appeal to 'intuition.' They can also feel free to ignore the history of mathematical research programs pursued by individuals alone in their studies or exchanging ideas in institutional settings. Cellucci notes that this view was shared by Hilbert, Russell, Tarski and Gödel, whose hopes were, of course, eventually dashed.

Cellucci also explores varieties of ampliative reasoning, comparing the analytic method as Plato understood it with the analytic-synthetic method as Aristotle understood it, and contrasting both with the analytic-synthetic method as 20th c. philosopher-logicians have construed it. Since new demonstrations change our understanding of the meaning of a problem and its objects, solving a problem is both a process of discovery and a process of justification. Thus, he argues, the analytic-synthetic process identified by, e.g., Hintikka and Remes in their rather logicist reading of Aristotle and Pappus in *The Method of Analysis* provides an impoverished account, according to which analysis should always be reversible into a deduction from already given axioms, or their deductive consequences, considered as premises (see Hintikka and Remes 1974). But then analysis would be carried out in a closed conceptual space, and once a proof was offered (which 'justified' the result) there would be no further need of other demonstrations.

Cellucci's careful account of analysis repays study. Most important, he makes clear that the hypotheses needed for solving a problem need not belong to the field of the problem, but may belong to other fields. The search for a solution to a problem is carried out in an open space (not the closed space delimited by a set of axioms and their deductive consequences). Mathematicians typically cut across what seem to be the boundaries of their own areas in search of solutions to problems. This is most strikingly true of number theory, which philosophers persist in regarding as safe inside the box of the Peano Postulates, despite those pesky incompleteness theorems; but number theorists, as I will show in the two chapters that follow, are always launching into the Beyond of topology, complex analysis, algebraic geometry, and category theory.

Cellucci also notes that the hypotheses for solving a problem are local, not global. Methods for solving problems are local: they depend on the peculiarities of the objects of the field and also on the peculiarities of the mathematical idioms in which they are couched. Thus we shouldn't be surprised that the formal languages of mathematical logic and set theory, designed to express the peculiar features of propositions on the one hand and sets on the other, are often not very useful for mathematicians who are trying to discover features of numbers, or Riemann surfaces, or matrices. Being local, not global, the hypotheses for solving a problem can be efficient. They can express and address the idiosyncracies of the problem, and exploit the specific strengths of the formal language. Problem searches try to reduce the search space, the domain within which the solution is sought, and this means using plausible hypotheses and exploiting known relations among terms (and among objects).

If we set these two features of the analytic method side by side, we can see a tension. The research mathematician is trying both to reduce the search space, but also to enlarge it: deep explanatory solutions to problems typically enlarge the setting of the original problem. Solutions to puzzles about integer solutions to certain simple polynomials are not solved by more and more computation, but rather by embedding the integers in other kinds of algebraic number fields (with help from complex analysis and group theory). On the one hand, good searches need to bridge domains; on the other hand, they depend on detailed knowledge of the domain in which the problem originates. But by the same token, searches depend on detailed knowledge of the domain brought in as an auxiliary, as well as a good acquaintance with abstract notions (like that of polynomial or group or vector space) that make the bridging possible. Those abstract notions, by the way, are not isolated concepts; they are more like a toolbox of instruments: the notion of group, for example, comes equipped with subgroup, coset, automorphism, permutation, homomorphism, quotient group, and so forth. Thus what we often find in mathematical problem-solving are patches from different fields juxtaposed or superimposed, to double or triple the information brought to bear on the problem, and reorganize our (local) understanding.

Cellucci is quite hostile to Frege and his project. "The second half of the 19th century marks the swan song of the view that logic must provide means to acquire knowledge. With the rise of mathematical logic, such a view is abandoned, and logic makes the deductivist turn… Frege is the key factor in this turn" (Cellucci 2013: 181). He criticizes Frege's basic assumptions about logic, his ideal of atomizing deduction, his view of mathematical practice in terms of the axiomatic method, his analysis of assertions and deduction, his foundationalism, and his logicist programme in general. He also considers the variant approaches to logic presented by Gentzen and Hilbert, but concludes that in each case, the high expectations were disappointed. Mathematical logic does not provide a means of discovery, a universal language for mathematics, a calculus of reasoning, a secure foundation for mathematics, or a discourse which is self-justifying. Moreover, by divorcing logic and method, it initiates a frivolous philosophical tradition of explicating discovery in terms of romantic genius, dreams, abnormal psychology and (sometimes) normal psychology.

Towards the end of *Rethinking Logic*, Cellucci develops his account of logic as encompassing methods for discovery as well as justification against the background of a 'naturalized epistemology' that draws broadly on the theory of evolution, anthropology, and current views about the biology of the human organism, especially in relation to neuroscience. And towards the end of the book, he examines 'rules of discovery,' including inductive reasoning, reasoning by analogy, generalization, specialization, metaphor and metonymy. These pages, I would argue, would have been better spent on studying the history of mathematics and the actual development of important recent problem-solutions. First, I don't think that the naturalist view that we should focus on plausibility instead of truth, and construe justification as always somewhat empirical and revisable, captures the peculiar nature of mathematical reasoning. The things of mathematics are determinate and unchanging as the things of nature are not: why would we ever revise our belief that the sum of the squares of the two oppositive sides of a Euclidean right triangle is equal to the square of its hypotenuse, or that $2 + 3 = 5$? We can always discover new aspects of numbers and figures when they are brought into relation with other mathematical items and systems. Cellucci is right that the process of analysis is unending, but not because mathematical knowledge is always revisable but because the objects of mathematics are infinitely inexhaustible. This claim becomes clear if we track the vicissitudes of a mathematical object over the course of a few centuries or millennia.

Second, as Cellucci himself observes, analytic methods of problem-solving are interestingly local. In between the alleged moments of irrational genius that bring insight (and about which logicist philosophers believe they have nothing to say) and the closed universality of the axiomatic method, lie the organized and principled and surprising local methods and bridging strategies of the mathematicians. We have to look carefully at the historical record, because the ability of mathematicians to deploy these strategies depends on a deep familiarity with the peculiarities of the objects and of the formal languages and schemata used to frame problems about them.

Carlo Cellucci's book bears a striking family resemblance to Danielle Macbeth's recent book *Realizing Reason: A Narrative of Truth and Knowing* (Macbeth 2014). Both books are the culmination of many decades of reflection upon, and refinement of, a complex view of logic, mathematics, knowledge and method. Both philosophers revisit the texts of Plato, Aristotle and Euclid continuing on to those of the present day, with significant sojourns in the early modern period and the 19th century. Both construe this history as a kind of dialectic, and want to revise our conception of reasoning to include not only a deductive logic of propositions (e.g. the predicate logic we all grew up with) but also an ampliative process of problem-solving and organization that employs a range of formal languages, including diagrams and iconic displays. Oddly, however, in Cellucci's story the villain is Frege, and in Macbeth's story Frege is the hero. Macbeth spends much of her book exploring the great mathematical flowering of European (especially German) mathematics, and explains how Frege's *Begriffsschrift* precisely captures the habits of mind that made it possible. Cellucci would argue that Frege's reformulation of logic made it precisely impossible to appreciate or capture those habits of mind. But what is philosophy if not dialectic?

At the beginning of *Realizing Reason*, Macbeth makes the following important observation: "…the symbolic languages of mathematics are quite unlike natural languages. Neither narrative nor sensory… they are special purpose instruments designed for particular purposes and useless for others. They are not constitutively social and historical, and they have no inherent tendency to change with use. Unlike natural languages, they also can (at least in some cases) be used merely mechanically, without understanding, used, that is, not as languages at all but as useful calculating devices" (Macbeth 2014: 108, and see also 58–68). Thus the communication of mathematical and scientific ideas to the general public is really very difficult, and it's often not clear what exactly is being communicated. What an equation means to a physicist may be quite different from what it means to a mathematician; what it means to a topologist may be different from what it means to a number theorist; and what it means to the general public, and how it means, may be something else altogether.

As Macbeth points out, while you can describe a procedure for finding (say) a complicated long division problem in English, you cannot calculate the result in English: you have to use Arabic numerals and the procedures you learned in primary school. Your calculation, written on paper with a pencil, doesn't describe or report the solution to the problem, but rather performs it—or one might say displays it or embodies it. You are reasoning *in* a system of signs, *in* a specially formulated artificial language. She notes further that whereas natural languages change with use and historical-cultural conditions, are first and foremost oral, and are amazingly versatile, formal-symbolic languages have no inherent tendency to change with time, are inherently written (we perform them by writing with them, correctly), and are designed for very specific uses. These special idioms make possible reasoning that has something like the rigor of deduction but is also ampliative, extending and adding to what we know (see also Serfati 2005; Mazur 2014).

Specialists in mathematics and science know how to *perform* mathematics or physics or chemistry: this is the point of all those damn problems at the end of chapters of textbooks that you just have to do to pass the exams and in fact to understand the subject matter. You have to learn to carry out each task in the special formal language designed for it, with the appropriate concepts, methods and procedures; mathematics and science (including the theoretical stuff, not just the experiments) must be performed. Thus, when a well-meaning scientist or mathematician describes a problem and the solution to the problem to us in English and in ordinary language, we don't really understand. In one sense we understand and in another sense we don't. A second-grader might nod her head when the teacher explains long division at the beginning of the year, but she doesn't really understand until, at the end of the year, she can perform it. At that point, she might say to her admiring, younger cousin, give me a big number and another big but smaller number, and I'll tell you what the answer is when you divide the first one by the second. And then too of course she has only taken the first steps down the long, long road of number theory, to understand the deep relations among natural numbers revealed by their decomposition as sums and on the other hand as products of other numbers.

Thus, as Margery Arent Safir remarks, in the Introduction to her edited volume *Storytelling in Science and Literature*, "Specialized material is made accessible to non-specialists only on condition of altering the language used," and, we might add, switching from performance to description (Safir 2014). A chemist reading an article in *Angewandte Chimie* is rehearsing the performance of another chemist which, in principle, he could himself re-perform, using the same technical language to describe what is going on and the same instruments and apparatus. An art historian, reading the same article to analyze how molecules are represented and admiring them for their symmetries, can only read it as a description, from the outside, and ask colleagues to define some of the technical vocabulary in plain English. But you can't perform chemistry, running reactions, synthesizing molecules and then breaking them up in order to study them by various methods, in plain English. You need the specialized languages of chemistry, the purified substances, the sheltered spaces, the instruments and the computers. For this very reason, the one episode in number theory that I try to explain 'all the way down,' in Chap. 4, includes an extensive glossary at the end, and a detailed account of proofs from an introductory textbook. I am trying to show how the proof is performed, not merely to describe it, in order to show what I need to show philosophically.

3 Kieran and David Egan: Teaching Complex Analysis

These reflections are important not only for philosophers, but also for teachers. To teach mathematics properly, we need to convey not just a set of facts, but also a repertoire of competences that can be performed (as we handle and employ mathematical idioms in order to solve problems) and so too a vivid sense of what real discovery in mathematics might be. Studying the history of mathematics can be illuminating here, for pedagogy can sometimes follow in the footsteps of earlier mathematicians. And philosophers also have something to learn in this pilgrimage.

High school students, like Leibniz and other mathematicians of the early modern era, are often puzzled by the complex numbers. What could we possibly mean by $\sqrt{-1}$? Why shouldn't we worry that the use of such a paradoxical concept might not tempt us into the pursuit of nonsense? (see Grosholz 1987). Historically, the existence and usefulness of complex numbers were not widely accepted until their geometrical interpretation around 1800, which was formulated at roughly the same time by Caspar Wessel, the Abbé Buée, Jean-Robert Argand and the great mathematician Carl Freidrich Gauss. (Leonhard Euler invented the notation $i = \sqrt{-1}$ late in his life, but it was Gauss' use of i in his *Disquisitiones arithmeticae* in 1801, which resulted in its widespread adoption.) Gauss published a memoir about the geometrical interpretation of complex numbers in 1832, which launched its wide acceptance in the mathematical world, aided also by the work of Augustin Louis Cauchy and Niels Henrik Abel (see Green 1976).

In the geometrical interpretation, every complex number is identified with an ordered pair of real numbers, (x, y)—which may also be written $x + iy$—and so further identified with the Euclidean plane. The notation of Cartesian coordinates for points on the plane (x, y) suggests itself here, where the number $z = x + iy$ is mapped onto the point $P = (x, y)$ with the real component x of z as abscissa, and the imaginary component y as ordinate. However, the notation of polar coordinates is also a fruitful way of writing complex numbers under this geometric interpretation, where r is the nonnegative length of the segment joining (x, y) to 0, and θ is the angle from the x-axis to this segment. We call $r = |z|$ the *absolute value* or *norm* or *modulus* of the complex number z and θ its *argument*. Indeed, this was Argand's mode of presentation of the complex numbers. The geometric interpretation, using this notation, immediately illuminates and is illuminated by Abraham de Moivre's formula (1730), because if we represent a complex number in this way, $z = |z|(\cos\theta + i\sin\theta)$, and if we know from de Moivre's formula that $(\cos\theta + i\sin\theta)^n = \cos(n\theta) + i\sin(n\theta)$, then we know that the absolute value of a product of complex numbers is the product of the absolute values of the factors, and the argument is the sum of the arguments of the factors. This means geometrically that complex multiplication corresponds to a dilation followed by a rotation. And the same insight holds for Euler's formula (1748), $e^{i\theta} = \cos\theta + i\sin\theta$, from which de Moivre's formula can be derived, as well as the equation we call Euler's identity, $e^{i\pi} + 1 = 0$, which in my experience never fails to enchant students of mathematics, at whatever level of study (see Nahin 2006/2011). A reader poll discussed in the *Mathematical Intelligencer* in 1990 named it as the most beautiful theorem in mathematics (see Wells 1990). For example, Euler's identity is in turn also a special case of another general identity that states that the complex nth roots of unity, for any n, add up to zero.

$$\sum_{k=0}^{n-1} e^{2\pi ik/n} = 0.$$

Students are delighted to discover that complex nth roots of unity can be found using trigonometry, and indeed geometry, because (as noted in Chap. 1) the complex nth roots of unity are the vertices of a regular polygon of n sides inscribed in the unit circle, $|z| = 1$ on the complex plane.

Thus the testimony of history is that we should introduce students to the complex numbers as they were introduced to the world of mathematicians between the mid-sixteenth century (when the Italians Ludovico Ferrari, Geronimo Cardan, Niccolo Tartaglia and Rafael Bombelli made important discoveries about the algebra they had inherited from medieval Latin and Arabic texts) and the mid-nineteenth century (when complex analysis, the theory of functions of a complex variable, flourished). In this period, we find a suite of modes of representation offered for the complex numbers, and when the geometric interpretation is offered, mathematical research on the complex numbers explodes. So we have an analogy, between historical sequence and pedagogical sequence: and how shall we understand its significance?

David Egan presents a kind of case study on a website hosted by the Imaginative Education Research Group (http://ierg.net), which offers advice about how to teach the complex numbers to high school students. The website includes a wealth of pedagogical materials, generated by a group of researchers at Simon Fraser University, in Vancouver, Canada, among whom Kieran Egan is perhaps the most visible. He is the author of *The Educated Mind: How Cognitive Tools Shape Our Understanding*, *Thinking Outside the Box*, and *Learning in Depth* (Egan 1999, 2007, 2011). His way of thinking about education is shaped by the writings of Ralph Waldo Emerson and John Dewey, as well as by the doctrines of Lev Vygotsky, whose work among Anglophone educators has recently enjoyed a resurgence, attested in the *Cambridge Companion to Vygotstky* (Daniels et al. 2007). One might mention in this regard the well-received book *The Gleam of Light: Moral Perfectionism and Education in Dewey and Emerson* by Naoko Saito (Kyoto University), written in the tradition of the Harvard neo-Pragmatists Israel Scheffler, Hilary Putnam and Stanley Cavell (Saito 2005). Here is one of the descriptions of their project, established in 2001 in the Faculty of Education at Simon Fraser University: "Imaginative Education is a way of teaching and learning that is based on engaging learners' imaginations. Imagination is the ability to think of what might be possible, in a manner that is not tightly constrained by the actual or taken-for-granted. It is the 'reaching out' feature of the mind, enabling us to go beyond what we have mastered so far. Without human imagination, no culture would look the way it does today, and no learner would be able to participate in and contribute to that culture" (see http://ierg.ca/about-us/what-is-imaginative-education/).

What does this research group mean by 'imagination,' a notoriously indeterminate member of the collection of faculties, as one looks back through the history of philosophy? The opening description on the website says that the group intends to build later forms of understanding on intellectual skills that are common in children in [traditional] cultures, such as story-telling, metaphor generation and recognition, image formation from spoken words, and so on. They reject the 'warehouse' model of learning, which depends on notions of storage and retrieval, and where the main challenge for the learner lies in mentally storing as much correct information as possible, and then being able to retrieve that information when needed. Instead, the research group seeks to promote educational methods that are not modeled on the assembly line or warehouse; so the topics that recur in their writings are developmental topics: somatic, mythic, romantic, philosophic and ironic understanding. The idea is that the ascent up these levels of understanding should not abandon but maintain and integrate the modes of education used on the lower levels (see Ruitenberg 2010).

On the website of the Imaginative Education Research Group, under Teacher Resources, there are lesson plans for a variety of scientific and mathematical topics, including among the latter differential calculus, decimalization, infinity, angles and complex numbers. The unit Complex Numbers, by David Egan, is aimed at 16 to 20-year-old students, for a unit that lasts 2–3 weeks (see http://ierg.ca/wp-content/

uploads/2014/01/Complex-Numbers-Unit-Plan.pdf). In Sect. 1, he identifies powerful underlying ideas: "It is very difficult to wrap the mind around the idea of imaginary numbers. Despite this bafflement, imaginary numbers fit quite sensibly into a system of complex numbers." Egan adds, "the experience of learning about complex numbers reinforces the tremendous power of abstract thinking, and the mathematical tools that facilitate it." That is, our earlier 'intuitive,' or 'visual' grasp of numbers must be revised and extended by the abstract tools of mathematical thinking. One of these tools is the Argand diagram; if we express complex numbers in the form $z = a + bi$, Egan notes, students will discover that the real numbers are actually exceptional; they are "the exceptional set of cases where $b = 0$" (Fig. 1). The teacher can also use this notation to introduce the Fundamental Theorem of Algebra: the complex numbers allow us, finally, to find ourselves in an algebraically closed number system.

Section 2 is entitled "Organizing the content into a theoretic structure." There Egan draws a distinction between a 'mathematical' and an 'intuitive' approach to numbers, and adds that an intuitive approach depends heavily on visualization, and on conceptualization in terms of concrete examples. Thus, we understand a negative number intuitively when we visualize it as the left-hand side of the real number line (left of zero), and conceptualize it as a bank account. By contrast, our arithmetic methods for multiplying large numbers one might call 'blind' conceptualization: though we can't 'see' what it means to multiply 435,678 by 963, 271 we can easily carry out the calculation on paper, using the wonderful positional method bequeathed to us by Indian and Arabic mathematicians, and we trust the results even though, he observes, we don't have the 'safety net' of intuition to check them.

In Sect. 2.2, concerned with organizing the unit, Egan asks "What meta-narrative provides a clear overall structure to the lesson or unit?" Here the teacher asks the students to wrestle with the problem of how to understand 5-dimensional Euclidean space; to see finally that they cannot visualize it; and then to resort to expressing it as $\mathbf{R}^5$, with points tagged by five real coordinates (v, w, x, y, z). They discover that although intuition has failed them, this notation that will allow them to formulate and solve problems about the geometry of 5-dimensional Euclidean space. And so

Fig. 1 Argand Diagram

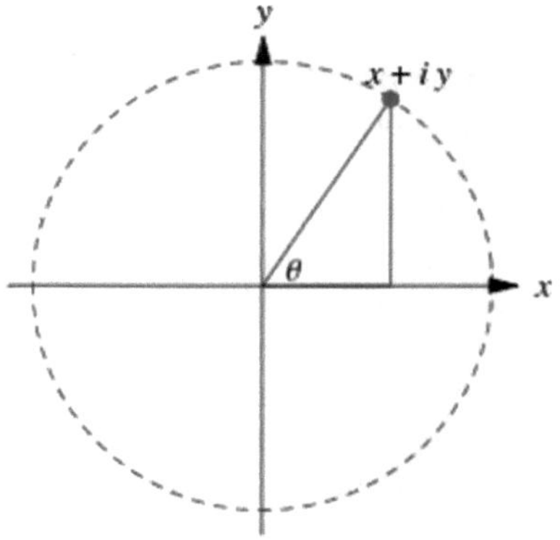

by analogy with complex numbers: the romance of this narrative, Egan surmises, is that mathematics can bravely venture where intuition fears to tread. And here he brings in the historical narrative: "A survey of the history of complex numbers shows the strong, and sometimes furious, opposition with which the idea of complex numbers was met… Many prominent mathematicians refused to accept complex numbers, and they only became widely accepted in the 19th century." Section 3 is entitled, "Developing the tools to analyze the theoretical structure." Here students learn to express complex numbers in the form $z = a + bi$, and then learn the rules for their addition, subtraction, multiplication, and division.

Egan observes that the Argand diagram makes complex numbers easier to visualize, and also provides a demonstration of the power of abstract, mathematical thinking: the one-dimensional number line has become a part of a 2-dimensional plane. (He adds that this is like the moment in a chess game, when a pawn is promoted to a queen.) The diagram can then be used to introduce the expression of complex numbers in polar coordinates, which might then lead to an introduction of their use in mechanics. This train of thought continues into Sect. 4, which continues the historical narrative, by pointing out that Gauss, at the age of 21, proved the Fundamental Theorem of Algebra: every algebraic equation has a solution in the set of complex numbers! This is also a good moment to point towards the world of complex numbers expressed as vectors using polar coordinates (using some linear algebra and trigonometry), and perhaps also to the world of functions of a complex variable.

Egan concludes, in the last three sections, that the point of the Complex Numbers unit is to encourage students to be bold, to reach out beyond their own 'intuitive' comfort zone to go beyond restrictions and limitations, "to navigate this fantastic world with the mathematical tools they have acquired." And he adds that the romantic approach works nicely with adolescents, who as we all know are keenly interested in voyages of exploration, like those of Odysseus and Robert Louis Stevenson. Thus too students can recognize retrospectively that the natural numbers, the integers, the rational numbers, the algebraic numbers, the reals and the complex numbers form a surprising but inevitable sequence, and that the complex numbers provide striking closure for that development.

Having criticized at length elsewhere the use of the term 'intuition' in the writings of Descartes, Kant and Brouwer, I will criticize its use in this lesson plan too (Grosholz 2007: Sects. 2.1 and 9.2). 20th c. mathematics was dominated by abstract algebra, and 20th c. philosophy of mathematics was dominated by logic. Thus it has become a commonplace to identify algebra and logic (and by association arithmetic) with mathematical reason, and geometry with 'intuition'. The narrative about the 19th century that many philosophers of mathematics favor is that geometric intuition led real and complex analysis astray into confusion and contradiction until Cauchy and Kronecker in one sense and Dedekind in another guided mathematicians out of the labyrinth through the arithmetization of analysis. While there is some truth to this particular myth, the other side of the story is that the use of geometry in most cases in 19th c. mathematics was *not* misleading and in many cases it was the key to important developments. Thus as we have seen the geometrization of complex numbers was essential to their acceptance and to the development of complex analysis; geometry provided the canonical examples that led to the formulation of group theory; and geometry, transformed by Riemann, lay at the heart of topology, which from the end of the 19th century throughout the 20th century in turn transformed much of modern mathematics.

In certain cases the demand for purity of method and the restriction of notation to one kind (motivated by the preference for ontological parsimony on the part of many philosophers, some mathematicians, and some educators) may be helpful for mathematical research and pedagogy. Sometimes if we investigate how much we can produce or how far we can proceed with very limited means, the outcome is mathematically suggestive. However, in many cases and indeed I believe in the most fruitful cases, we find mathematicians juxtaposing and even superimposing a variety of notations or more generally 'modes of representation,' that is, we find them multiplying rather than restricting their 'paper tools,' as Ursula Klein calls them (see Klein 2001).

In order to counter and clarify the various uses of the term 'intuition' as opposed to reason, I prefer to use the terms 'iconic' and 'symbolic,' borrowed from the American philosopher C. S. Peirce. Some mathematical modes of representation are iconic, that is, they picture and resemble what they picture; others are symbolic and represent by convention, 'blindly' and without much resemblance (Grosholz 2007: 4). As Macbeth shows, we 'perform' mathematics in both iconic and symbolic idioms. In many cases, problems in mathematics are most successfully understood, addressed and solved when the problematic things that give rise to them are represented by a consortium of modes of representation, some iconic and some symbolic. Both kinds do important conceptual work: symbols typically help to analyze and distinguish, and icons help to unify and stabilize reference (though not always, as we shall see). We need to do both at the same time in order to identify, reformulate and solve problems.

Thus, as I see it, the reason why the geometric interpretation of complex numbers moved mathematical research forward historically and why it aids students pedagogically is because it gives us a repertoire of modes of representation that can be used in concert to understand what complex numbers are and how to use them. Students are not leaving behind the timid formulation of i as the square root of (-1), or as the solution to the equation $x^2 + 1 = 0$ given in the original context of an algebra of arithmetic transformed by its use in analytic geometry. Rather, they are using it together with the Argand diagram, which is no more or less 'intuitive' or 'mathematical' than the algebraic representation. But it is certainly more iconic and spatial, whereas the algebraic representation is more 'blind' and symbolic. The Argand diagram suggests as well two different symbolic formulations, one in Cartesian and one in polar coordinates; the latter immediately brings in the notation of trigonometry and the transcendental functions, originally so foreign to analytic geometry, that underlie it. This combination of modes of representation, as we have seen, reveals our old friend the circle in an entirely new way, as the home and factory of the sine and cosine functions. All of mechanics, we might say, lies folded up in the circle.

So I would re-write the romance of complex numbers. What we find historically is the *addition* of an important iconic, geometric representation of complex numbers to the existing algebraic representation, which in turn suggests a trigonometric representation that lends itself to mechanics. In the end, we'll have at least four modes of representation on the page, to help us think through problems! We could call this a kind of laboratory work, investigating mathematical things on the combinatorial space of the page (as Jean Cavaillès called it) with paper tools. So what we are teaching students is not how to leave behind the 'intuitive' for the 'mathematical,' but rather how to profit from and think together a new range of representations, all of them mathematical, some of them iconic and some of them symbolic, in order to investigate complex numbers and complex functions more effectively. The pedagogical point is still innovation, critical thinking and originality, and the goal is teaching students how to make conceptual breakthroughs. But now, I would argue, we are guided by a more accurate reading of history, and by philosophical ideas that do not promote algebra and logic at the expense of geometry.

References

Cellucci, C. (1998). *Le ragioni della logica*. Rome and Bari: Laterza & Figli.
Cellucci, C. (2002). *Filosofia e matematica*. Rome and Bari: Laterza & Figli.
Cellucci, C. (2007). *La filosofia della matematica del Novecento*. Rome and Bari: Laterza & Figli.
Cellucci, C. (2008). *Perché ancora la filosofia*. Rome and Bari: Laterza & Figli.
Cellucci, C. (2013). *Rethinking logic: Logic in relation to mathematics, evolution and method*. Dordrecht: Springer.
Chemla, K., & Shuchun, G. (Tr. and Eds.). (2004). *Les neuf chapitres: La classique mathématique de la Chine ancienne et ses commentaires*. Paris: Dunod.
Daniels, H., Cole, M., & Wertsch, J. V. (Eds.). (2007). *Cambridge Companion to Vygotsky*. Cambridge: Cambridge University Press
Debuiche, V. (2013a). Perspective in Leibniz's Invention of Characteristica Geometrica: The problem of Desargues' influence. *In Historia Mathematica, 40*, 359–385.
Debuiche, V. (2013b). L'expression leibnizienne et ses modèles mathématiques. *Journal of the History of Philosophy, 51*(3), 409–439.
Egan, K. (1999). *The educated mind: How cognitive tools shape our understanding*. Chicago: University of Chicago Press.
Egan, K. (2007). *Thinking outside the box*. New York: Teachers' College Press.
Egan, K. (2011). *Learning in depth*. Chicago: University of Chicago Press.
Green, D. R. (1976). The historical development of complex numbers. *The Mathematical Gazette, 60/412*, 99–107.
Grosholz, E. (1987). Two Leibnizian manuscripts of 1690 concerning differential equations. *Historia Mathematica, 14*, 1–37.
Grosholz, E. (1991). *Cartesian method and the problem of reduction*. Oxford: Oxford University Press.

Grosholz, E. (2005). Lessons about method from a Chinese mathematical classic. Review of K. Chemla & G. Shuchun (Tr. and Eds.), *Les neuf chapitres*. In *Gazette des Mathématiciens*, *105*, 49–56.

Grosholz, E. (2007). *Representation and productive ambiguity in mathematics and the sciences*. Oxford: Oxford University Press.

Hintikka, J., & Remes, U. (1974). *The method of analysis: Its geometrical origin and its general significance*. Stuttgart: Springer.

Klein, U. (2001). *Tools and modes of representation in the laboratory sciences*. Dordrecht: Kluwer.

Leibniz, G. W. (1682). De vera proportione circuli ad quadratum circumscriptum in numeris rationalibus expressa. *Acta Eruditorum, Feb. 1682*; reprinted in Leibniz 1978 V: 118–22.

Leibniz, G. W. (1692). De la chainette. *Journal des Scavans 1692*; reprinted in Leibniz 1978 V: 258–263.

Leibniz, G. W. (1962). In C. I. Gerhardt (Ed.), *Die Philosophischen Schriften* (Vol. I–VII) Hildesheim: Georg Olms.

Leibniz, G. W. (1978). In C. I. Gerhardt (Ed.), *Mathmatischen Schriften* (Vol. I–VII) Hildesheim: Georg Olms.

Leibniz, G. W. (1989). In M. Parmentier (Tr.), *La naissance du calcul différentiel: 26 articles des* Acta Eruditorum. Paris: Vrin.

Leibniz, G. W. (1990). In E. Knobloch & W. Contro (Eds.), *Mathematische Schriften, Geometrie – Zahlentheorie – Algebra, 1672–1676* (Series VII, Vol. 1). Berlin: Akademie Verlag.

Leibniz, G. W. (1993). In E. Knobloch (Ed.), *De quadratura arithmetica circuli ellipseos et hyperbolae cujus corollarium est trigonometria sine tabulis*. Göttingen: Vandenhoeck & Ruprecht.

Leibniz, G. W. (2004). M. Parmentier (Tr.), *Quadratura arithmétique du cercle, de l'ellipse et de l'hyperbole et la trigonométrie sans tables trigonométriques qui en est le corollaire*. Paris: Vrin.

Macbeth, D. (2014). *Realizing reason: A narrative of truth and knowing*. Oxford: Oxford University Press.

Mazur, J. (2014). *Enlightening symbols: A short history of mathematical notation and its hidden powers*. Princeton: Princeton University Press.

Nahin, P. J. (2006/2011). *Dr. Euler's Fabulous Formula*. Princeton: Princeton University Press.

Ruitenberg, C. (Ed.). (2010). *What do philosophers of education do?*. Oxford: Oxford University Press.

Safir, M. (Ed.). (2014). *Storytelling in science and literature*. Lewisburg: Bucknell University Press.

Saito, N. (2005). *The Gleam of light: Moral perfectionism and education in Dewey and Emerson*. New York: Fordham University Press.

Serfati, M. (2005). *La révolution symbolique: La constitution de l'écriture symbolique mathématique*. Paris: Editions Pétra.

Vuillemin, J. (1997). La Question de savoir s'il existe des réalités mathématiques a-t-elle un sens? *Philosophia scientiæ 2/2*: *Actes du Colloque international Nelson Goodman, Pont-à-Mousson*, *2*, 275–312.

Wells, D. (1990). Are these the most beautiful? *The Mathematical Intelligencer, 12*(3), 38–41.

Chapter 4
Algebraic Number Theory and the Complex Plane

1 Early Problems in Number Theory

If we are interested in number theory and have been trained in a certain philosophical tradition, the pertinence of complex analysis to number theory will be puzzling. The logicist and formalist tradition, even modified as it must be now by the meta-theorems of mathematical logic, often seems to claim that all of classical pure mathematics can be deduced from the theory of sets adumbrated by predicate logic (hopefully confined to first and second order predicate logic). The reductionist account involves a sequence of theory reductions, beginning with the reduction of reasoning about the integers and the rational numbers to first-order Peano Arithmetic. If we start with a first-order theory of Peano Arithmetic, for example, we assume that we have at our disposal the natural numbers; they are 'given' as the intended model of the theory. However, it is not clear exactly how they are given. They are not given by the formal theory, which is not categorical and cannot by itself pick out the intended model from a spectrum of non-standard models which are nevertheless isomorphic to the natural numbers from the point of view of the theory. So we have to bracket the question of how we know they are available.

We suppose then that 'the language of number theory' is deducible from the axioms of Peano Arithmetic; we can adopt Enderton's preferred formulation A_E presented in his *A Mathematical Introduction to Logic* (Enderton 1972: 193–194). It is expressed in first-order predicate logic with the addition of a constant symbol **0**; a one-place function symbol **S** (the successor function); a two-place predicate symbol that denotes the strict linear ordering of the natural numbers <; and finally +, •, and E, two-place function symbols intended to denote the operations of addition, multiplication and exponentiation. The terms of this language thus include the sequence **0**, **S0**, **SS0**, **SSS0**… et cetera; also included in this language are arithmetic truths like **S0** < **SS0**, and **S0** + **SS0** = **SSS0**, and so forth, and logical compounds involving individual and n-place predicate symbols, and formed with negation, the operations 'and' and 'or,' and the adjunction of quantifiers. (Enderton

E.R. Grosholz, *Starry Reckoning: Reference and Analysis in Mathematics and Cosmology*, Studies in Applied Philosophy, Epistemology and Rational Ethics 30, DOI 10.1007/978-3-319-46690-3_4

1972: Chap. 3, and see 68 and 174, where Enderton finesses the status of the equality sign—it is neither a part of pure predicate logic nor is it part of the special vocabulary of arithmentic.) If we begin with the example of the counting numbers, we notice the simple way in which they are generated, adding a unit to the unit, and then adding another unit to the sum of the unit and a unit, and then adding another unit *et cetera,* ad infinitum: the counting numbers seem to arrive with their rational content in full display.

However, early on (in human history and in the education of children) the counting numbers lead to patterns that seem mysterious. The natural numbers **N** can be divided into odd and even numbers; it is easy to present, e.g., the odd numbers, to find the next one in the list, and to give a general rule that captures all of them. But **N** can also be divided into prime and non-prime numbers; it is not so easy to present this list by finding the next and the next and the next one by using, for example, the sieve of Eratosthenes. And there is no formula that captures them all, as $2n$ and $(2n + 1)$ capture the even and odd numbers. This difference stems from the difference between the trivial additive decomposition of the natural numbers (which logicians make much of) and the highly non-trivial multiplicative decomposition of the natural numbers (which logicians try to finesse: the primes after all are a **recursive set**). In the historical development of number theory, however, insight is not arrived at merely by combining what is "already there," like facts of addition and multiplication. Going beyond **N**, in the variety of ways in which one can go beyond **N**, is the only way to discover why the odd patterns evident in **N** emerge, to understand the deeper reasons for them. This process leads to novel investigations of the fine structures inherent in **N**, which are themselves not discoverable until mathematicians have brought some of those external structures to bear on **N**; it also changes the meaning of basic notions like number, prime and unit. This "going beyond" includes bringing **N** into relation with Euclidean geometry; with the integral domain of integers **Z**, the ring of rationals **Q**, the field of reals **R**, and the field of complex numbers **C**; with various algebraic number fields like **Q**[i] and other **cyclotomic fields**, or **Q**[$\sqrt{2}$] and other quadratic fields; with the fields of ***p*-adic numbers**; and with the complex plane as the staging ground for complex analytic or topological methods.

In the 17th century, Fermat noted that there were many positive integral solutions to the equation $x^2 + y^2 = z^2$, but none to be found for the equations $x^3 + y^3 = z^3$ or $x^4 + y^4 = z^4$ or $x^5 + y^5 = z^5$. How did he see this as a pattern in the first place, and why did he find it puzzling and in need of explanation? One important bit of background knowledge that set the stage for this perception was the Pythagorean Theorem. This result from classical antiquity, as we all know, associated numbers with a certain geometrical figure (the right triangle) in a deep and illuminating way: for every right triangle, the square of the length of the hypotenuse is equal to the sum of the squares of the lengths of the two legs, that is, the two sides opposite the hypotenuse. Examples of triplets of positive whole numbers that in this case satisfy both the constraints of arithmetic (the multiplication and addition tables) and the constraints of geometry (the nature of a triangle in Euclidean geometry and the constraints put on a triangle by its being right) were known to the Pythagoreans: 3, 4, 5; 5, 12, 13; 8, 15, 17; even 119, 120, 169.

However, Fermat noted *both* the existence of these triplet-solutions and their absence in the case of analogous equations with exponents of higher degree than 2. Why did he notice this absence? And why did he perceive this absence as anomalous or puzzling? Why hadn't this pattern bothered anyone else before? One answer to these questions is that Fermat, like Descartes, used the notation of algebra, where variables represent suites of numbers in one sense and indeterminate numbers in another sense, and constants represent distinguished, determinate numbers (so variables and constants play different conceptual roles), and equations replace proportions. Thus he could consider an expression like $x^2 + y^2 = z^2$, and ask what might happen if one replaced 2 with another integer. The polynomial itself becomes a conceptual display and an object of mathematical investigation. Descartes discovered an analogous puzzle (in this case among curves rather than numbers) apropos Pappus' Problem: after he finished solving the problem for the conic sections (that is, for curves associated with quadratic equations in two variables), he posed the problem hypothetically for cubic equations and analogous equations of higher degree. Like Fermat, he was puzzled by the difficulty of obtaining solutions and discerning a taxonomy even for cubic curves, and like Fermat he boasted unrealistically that nonetheless his methods would lead straightway to the solutions sought (see Grosholz 1991: Chaps. 1 and 2).

One reason why the background knowledge of Fermat and Descartes led them to perceive these facts (there seem to be no solutions—among the integers **Z**—to the equation $x^n + y^n = z^n$ when n is equal to 3 or 4 or 5 or higher; and, it is very hard to identify and classify all the cubic curves, and even harder to identify and classify the higher algebraic curves of n degree) as *puzzling*, is that they understood them against the background of arithmetic, that is, in light of the method of mathematical induction. That method shows that if you can prove a claim involving n for the first case 1, and then prove that if the case involving n holds, the case $n + 1$ must hold, then you have proved your claim for every n. But this method only works when n indexes the cases in a certain way, and when the relation between case n and case $n + 1$ is a certain kind of relation. (For example, once arithmetic has been allied with geometry in the way that algebra facilitates, indexing by n becomes really problematic because increasing the dimension of Euclidean space increases the complexity of that space and the items in it and their mutual relations in a way that is serious, if not dire.) Thus, it seemed to Fermat that if $x^2 + y^2 = z^2$ had lots of positive integral solutions, $x^3 + y^3 = z^3$ should have them too. But it didn't. Background knowledge of arithmetic, and the yoking of arithmetic and geometry by the new algebra, made this fact show up as anomalous, as a rupture in the intelligibility of mathematics, as requiring an *explanation*.

An explanation of this puzzling fact could not be pursued by mere induction, that is, by computing more examples of triplets or by looking harder for examples of the missing triplets. The explanation lay in three insights, well articulated only in the 19th century. The first is that the natural numbers can be better understood if they are embedded in other number systems that are larger and may have different kinds of systematic features. The second is that the natural numbers have a fine structure, or rather a repertoire of fine structures, which only becomes apparent in light of these embeddings. The third, twentieth century insight is that there are deep correlations between the repertoire of fine structures and the greater embeddings: this is the central insight of Class Field Theory. As I go through these three developments in this chapter, I will endeavor to exhibit the novel concept clusters, and show how they arise from the exploitation of structural ambiguities, which bring together ideas that formerly did not interact at all, in the service of problem-solving: not mere proof, but cogent, systematic explanation (see Hicks 2013 for important insights into the formation of concept clusters). But first, I want to review the mid-twentieth century account of theory reduction, in order afterwards to show that the successful explanation of these facts of number theory cannot be captured by that account.

2 Butterfield's Critical Assessment of Nagel's Account of Theory Reduction

In an essay on the topic of time, the British philosopher of science Jeremy Butterfield offers a useful discussion of the philosophical problem of reduction. (Butterfield and Isham 1998/1999). He begins by reviewing the standard account of theory reduction proposed by Ernst Nagel in his influential book, *The Structure of Science* (1961) (Nagel 1961). Nagel's well-known version of theory reduction proposes that a theory T_1 is reduced to another theory T_2 when the theorems of T_1 are a subset of those of T_2, so that T_1 is a subtheory of T_2. Butterfield observes, "However, one needs to avoid confusion that can arise from the same predicate (or other non-logical symbol) occurring in both theories, but with different intended interpretations. This is usually addressed by taking the theories to have disjoint non-logical vocabularies. Then one defines T_1 to be a *definitional extension* of T_2, if and only if one can add to T_2 a definition of each of the non-logical symbols of T_1 in such a way that T_1 becomes a sub-theory of T_2. That is: in T_2, once augmented with the definitions, we can prove every theorem of T_1. (The definitions must of course be judiciously chosen, with a view to securing the theorems of T_1.)" (Butterfield and Isham 1998: 6). He then raises the following interesting question: "which operations for compounding predicates, and perhaps other symbols, does one allow oneself in building the definitions of the terms of T_1? It has been usual to consider a very meagre stock of operations, viz. just the logical operations: the Boolean operations of conjunction and negation, and the application of the quantifiers 'all' and 'some'" (Butterfield and Isham 1998: 7). For theory reduction to be plausible in physics, he argues, this stock of operations should include as well other standard operations like taking derivatives, integrals, orthocomplements, completions, and so forth. Physical theories, he notes, use mathematical apparatus that is 'high up' in the deductive chain relative to basic logic and set theory, like calculus, so that in these cases the 'underlying logic' must be much stronger. This limitation of the standard account of theory reduction was registered even by its proponents. "Nagel adds to the core idea of definitional extension some informal conditions, mainly motivated by the idea that the reducing theory should explain the reduced theory" (Butterfield and Isham 1998: 7).

Butterfield observes that this means definitional extension is at once too weak and too strong to capture what actually happens in scientific practice. It is too weak because, even if T_1 is a definitional extension of T_2 so that the extension of each of its predicates has the same extension as a compound predicate built from T_2's vocabulary, "there may well be aspects of T_1, crucial to its functioning as a scientific theory, that are not encompassed by (are not part of) the corresponding aspects of T_2" (Butterfield and Isham 1998: 9). The properties of T_1 might simply be different from those of T_2, even the compound properties of T_2; or T_1 might have aspects to do with explanation, modelling or heuristics that are not encompassed by T_2. These aspects of T_1 that 'outstrip' T_2 vary, Butterfield observes, from case to case; the controversy that ensued over how to supplement definitional extension after Nagel proposed it shows that there is no general consensus and therefore no way to formalize this supplementation, widely recognized, even by Nagel, to be required. Thus, although physics offers many examples of theory reduction and definitional extension, this formal pairing typically requires further 'supplementation' (to capture the ways in which T_1 outstrips T_2) and such supplementation varies from case to case. Butterfield concludes, "there may well be no single 'best' concept of reduction—no 'essence' of reduction," that is, no formalizable supplement (Butterfield and Isham 1998: 10).

Definitional extension is also too strong. Nagel's proposal is widely regarded as too strict; there are many cases of useful reductions in science that don't live up to his standards, for example, cases where T_2 reduces T_1 while remaining inconsistent with it, as Feyerabend argued (Feyerabend 1962/1981). Newton's reduction of the principles of Kepler, Galileo and Descartes are taken to be an illustration of this point. Reduction often involves approximation, a point conceded even by Nagel. "More generally, various authors have suggested ... that reduction often involves T_2 including some sort of analogue, T^* say, of T_1. They require this analogue to be close enough to T_1, in such matters as its theoretical properties and the postulates concerning them, and/or its explanatory resources, and/or its observational consequences, that one is happy to say that 'T_2 reduces T_1'" rather than merely concluding that the theories are incommensurable (Butterfield and Isham 1998: 11). Here again, controversy continues about what conditions are required of T^* so that it is 'close enough' to T_1, and so once again the possibility of formalizing some 'essence' of reduction seems elusive; and some conceptual and explanatory disparity remains. Thus there seems to be no way to draw a sharp (formalizable) line between reduction, and emergence or supervenience or replacement or incommensurability, in the physical sciences.

Oddly enough, however, Butterfield begins the discussion by claiming that the overblown optimism shown by mid-20th c. philosophers of science about the possibility of formalizing definitional extension for physics stems from "the striking success of studies in logic and mathematics, from the mid-nineteenth century onwards, in showing various pure mathematical theories to be definitional extensions in the above sense, i.e., using just these logical operations, of others. Indeed, by concatenating such deductions with judiciously chosen definitions, one shows in effect that all of classical pure mathematics can be deduced from the theory of sets." (Butterfield and Isham 1998: 6). Presumably by this claim he has in mind the alleged reduction of the natural numbers to sets, the reduction of the integers and rationals to natural numbers, the reduction of the reals to the rationals, and the reduction of geometry to the numerical models of analytic geometry and real analysis. Yet none of these reductions is straightforward; each must be understood variously in syntactical and semantic terms; and each is either too strong or too weak, to use Butterfield's vocabulary, when measured against the ideal reduction of Nagel. Here I argue that Butterfield's arguments about the failure of philosophers to formalize definitional extension in physics apply just as well, and for the same reasons, to mathematics and specifically to number theory. While the reductions just listed are important and fruitful, taken together they are a motley and do not amount to an example of Nagelian reduction.

3 Theory Reduction Applied to Number Theory and Arithmetic

Is it true that all of classical pure mathematics can be deduced from the theory of sets? This claim must be spelled out in detail so that the sequence of theory reductions can be assessed. Let us look first at what should be the most straightforward case, the reduction of reasoning about the integers and the rational numbers to first-order Peano Arithmetic. If we start with a first-order theory of Peano Arithmetic, for example, we assume that we have at our disposal the natural numbers; they are 'given' as the intended model of the theory. As noted above, the formal theory is not categorical and cannot by itself pick out the intended model from a spectrum of non-standard models which are nevertheless isomorphic to the natural numbers from the point of view of the theory. So there is an unresolved problem already concerning how we know the natural numbers are available.

We suppose then that 'the language of number theory' is deducible from the axioms of Peano Arithmetic, as formulated in Enderton's textbook and cited at the beginning of this chapter. An arithmetical set is a set of natural numbers (supposing that they are available) that can be defined by a well-formed formula of first-order Peano Arithmetic, A_E. There is a hierarchy of arithmetical sets, whose levels are registered by the number and order of universal and existential quantifiers that precede the formulae that live on each level. The first level consists of the recursive sets (represented by formulae with no quantifiers), and the second consists of the recursively enumerable sets, represented by formulae preceded by a finite number of existential quantifiers. The rest of the hierarchy is produced by adding blocks of existential and universal quantifiers, so that the increasing internal complexity of the formulae reflects the increasing complexity (or, decreasing computability) of the sets of natural numbers. This is why quantifier elimination is an important tool in model theory, so that formulae can be re-written in a way that exhibits their logical complexity precisely. Any arbitrary countable set of natural numbers can be located somewhere in the arithmetical hierarchy. Thus, the integers, the set of n-tuples of integers for a given n, and the rational numbers can all be shown to be arithmetical sets, if we use the trick of Gödel numbering to index formulae that define each element of a given set, and then check to see if the set of corresponding Gödel numbers (which are all natural numbers) is arithmetical. As noted above, the set of prime numbers turns out to be recursive, even though it is increasingly difficult to use the 'sieve of Eratasthenes' for larger and larger primes. Thus we can say that a definitional extension that exploits Gödel numbering allows the set of primes as well as the set of integers to be defined within Peano Arithmetic.

However, definitional extension is inherently ampliative. The strategy of Gödel numbering depends on the assumption that the formal language is sandwiched in between schoolroom arithmetic and a meta-language which Enderton characterizes as 'English.' Schoolroom arithmetic is expressed in Indo-Arabic numerals with algebraic conventions developed in the 17th century; it is not a formalized language, but rather the expression in a variety of formal and informal idioms of a collection of interesting facts, solved and unsolved problems, and procedures for calculation. Moreover, its boundaries are not clear: does the problem, to show that an odd prime p can be written as the sum of two squares (that is, $x^2 + y^2$ where x and y are integers) if and only if $p \equiv 1 \pmod 4$, belong to schoolroom arithmetic? (I examine this problem later in this chapter.) And English is certainly not a formalized language. Fragments of schoolroom arithmetic and of English are formalizable in the language of first-order predicate logic, as every student in an introductory logic class discovers; but this local translatability does not make either arithmetic or English an object language or a meta-language in the technical sense intended by logicians. The pretense that the relations among schoolroom arithmetic (not to speak of modern number theory!), A_E, and English are precisely the relations among a formal object language, a formal language, and a formal meta-language is motivated by the logicizing philosopher's wish to make the relations among these 'theories' resemble the mid-twentieth century model of theory reduction, where the dictionaries are trivial. But they never are.

Let me underscore the point by quoting from Enderton's *A Mathematical Introduction to Logic*, Chap. 3, 'Undecidability.' He introduces the language of number theory, 'a first-order language with equality and with the following parameters…' and then adds, "We will let R be the intended structure for this language. Thus we may informally write R = (**N**, **0**, ***S***, <, +, •, ***E***). (More precisely, $| \mathrm{R} |$ = N…, O^{R} = 0, etc.) By number theory we mean the theory of this structure, ThR" (Enderton 1972: 174–175). He adds a bit later that ThR is "a very strong theory and is neither decidable nor axiomatizable," so that we should concern ourselves with A_E (the axiom set he proposes on pages 193–194, as noted above, along with all its theorems). In any case, by treating schoolroom arithmetic 'informally' as if it were 'the intended structure,' Enderton slips in the assumption that it is already structured as an object theory, that reference to the natural numbers can be treated as if it were a map from one discourse (A_E) to another discourse (R) and that the natural numbers are available as a set which is ambiguously part of R, over which the universal quantifier of A_E can range. But schoolroom arithmetic is not structured like a theory and it is not axiomatized: the whole point of the Peano Axioms was to axiomatize it! It is a discourse, but as a discourse it is a messy collection of facts, solved and unsolved problems, and procedures for calculation and 'proof by recurrence,' dependent on the additive decomposition of the natural numbers as sums of units, and their multiplicative unique prime decomposition. And it refers to the natural numbers (using Indo-Arabic notation), which are not themselves discourse even if our access to them is through discourse, and which cannot without further—ampliative and contestable—argument be construed as a set.

Here are Enderton's remarks that introduce the Sect. 3.4, 'Arithmetization of Syntax.' "(1) Certain assertions about wffs can be converted into assertions about natural numbers (by assigning numbers to expressions). (2) These (English) assertions about natural numbers can in some cases be translated into the formal language ..." (Enderton 1972: 217). But English is not a meta-language any more than arithmetic is a structure, model or object-language. Moreover, the dictionary supplied by Enderton, which maps A_E to R (the 'intended structure' for A_E, which is in fact schoolroom arithmetic) is not part of A_E, and neither is the dictionary, or rather code manual, of Gödel numbering which maps English to A_E, part of A_E. They are instead good examples of sideways ampliative reasoning.

Nagel and Newman, in Chap. 7 of *Gödel's Proof*, write "Gödel described a formalized calculus within which all the customary arithmetical notations can be expressed and familiar arithmetical relations established," a formalized language like A_E. "Gödel first showed that it is possible to assign a unique number to each elementary sign, each formula (or sequence of signs), and each proof (or finite sequence of formulas). This number, which serves as a distinctive tag or label, is called the "Gödel number" of the sign, formula, or proof." (Nagel and Newman 1958: 68–69). Nagel and Newman spend the next few pages (69–76) spelling out what the assignment looks like in detail, concluding that this set of directions establishes a one-one correspondence between expressions in the formalized language and certain natural numbers. In virtue of the unique prime decomposition enjoyed by the natural numbers, each term, wff (well formed formula) and sequence of wffs, can be retrieved from the natural number that codes it by unpacking that number's prime decomposition. But where does this assignment and unpacking take place? Not, clearly, in A_E. They observe that Gödel "showed that all meta-mathematical statements about the structural properties of expressions in the calculus can be adequately mirrored within the calculus itself... Since every expression in the calculus is associated with a Gödel number, a meta-mathematical statement about expressions and their relations to one another may be construed as a statement about the corresponding (Gödel numbers and their arithmetical relations to one another.") (Nagel and Newman 1958: 76–77).

What is this meta-mathematics? Nagel and Newman characterize it as expressing "the structural properties of expressions in the calculus" and "relations of logical dependence between meta-mathematical statements," as if it were, not English, but rather a slightly aggrandized version of A_E, another formal language that could refer to the wffs of A_E, that would be about the wffs of A_E, and say modestly formal things about them, say A_E'. (Recall that A_E is certainly about the discourse of arithmetic and is also sometimes alleged to be about the natural numbers, which are themselves not a discourse.) But the meta-language must do much more than that. Perhaps there is a way to say in A_E' that one formula b of A_E is an initial part of another formula a of A_E, or that a certain sequence of wffs x in A_E is a proof of a certain wff z in A_E. And perhaps there is a way to say in A_E that b is a factor of a (in other words, that 'is a factor of' can be suitably defined in A_E) or that x and z are related by the arithmetical formula 'Dem (x, z).' (For an explanation of Dem (x, z) see Chap. 5.) But in the meta-language required by the demands of Gödel's proof, we must also be able to say that "the smaller formula '(pvp)' can be an initial part of the axiom '(pvp) → p' iff the (Gödel) number b, representing the former, is a factor of the (Gödel) number a, representing the latter," and that "the sequence of formulae with Gödel number x is a proof of the formula with Gödel number z iff Dem (x, z) is true." The cogency and meaningfulness of both instances of "iff" depends on the intervention of the dictionary in the argument: we have invoked Gödel numbering. But the dictionary belongs neither to A_E nor to A_E': it requires us to correlate terms and wffs (what A_E' is about) with numbers (what A_E is about), and, as Nagel and Newman argue at length in a footnote, a numeral is a sign (a term) and a number, in particular a Gödel number, is something a term designates: "we cannot literally substitute a number for a sign." (Nagel and Newman 1958: 82–84). This important distinction, they explain, stems from the difference between mathematics and meta-mathematics.

I want to make two points about the failure of the relation between A_E and R and the relation between A_E' and A_E to live up to the standards of ideal theory reduction. First, the mismatch is useful. Some of the amplification of knowledge that takes place between A_E and R is contained in the model theoretic insights about Wiles' proof that I discuss in Chap. 5. The amplification that takes place between A_E' and A_E includes the substance of Gödel's proof and the meta-theoretic results about formal languages that follow in its wake. The demands of pure theory reduction are too strong, because when we go up from arithmetic to A_E and then to A_E' we lose the specificity of the natural numbers, but we can work with an analogue that is 'good enough' and that reveals the logical complexity of reasoning in arithmetic as well as novel features of the languages of logic; so we don't reject the reduction as mere incommensurability. And the reduction is too weak: the natural numbers re-represented as 0, S0, SS0, … may be useful for logic but this idiom makes ordinary arithmetic impossible. Arithmetic, as Macbeth would observe, cannot be performed in this idiom. Moreover, there are important aspects of the natural numbers, crucial to their functioning within the series of problems that characterize 18th through 21st century number theory, which are not encompassed by and are not part of a system of terms and wffs. The properties of natural numbers are simply different from those of logical terms and wffs, requiring different idioms and methods, and those distinctive features play a crucial role in the explanations, modeling and heuristics of number theory. For example, the natural numbers are the only mathematical items that index, and count, themselves; that is why Gödel must use them to index the pertinent terms and wffs. The natural numbers enjoy unique prime (multiplicative) decomposition; but there is no such thing as a prime term or logical formula (or, for that matter, a prime set). The concept of prime does not arise in the consideration of sets or logical formulae; consider a collection of sets or formulae as long as you like, but you will never discern any "primes" among them.

Similarly, the reduction supposed to hold between the rational numbers and the natural numbers within A_E is too strong; specifically, we do not retrieve the rational numbers here, but only an equivalence class of isomorphic systems identified by the pertinent set of formulae in the arithmetic hierarchy. Recall that we cannot retrieve the natural numbers from A_E because it is not categorical; we have to sacrifice the power of second order ThR for the good behavior of first order A_E. Second, the reduction is too weak; there are aspects of the rational numbers crucial to their functioning in mathematical problems that are not encompassed by their re-representation as a set of Gödel numbers located somewhere in the lower reaches of the set-theoretical hierarchy of sets of natural numbers, associated with a formula in the logical arithmetic hierarchy. These aspects include their important topological features, when we think of them as points in the continuum of the line: unlike the discrete natural numbers they are dense, and unlike the real numbers they are not complete. Considered by themselves as a topological space, they are the unique countable metrizable space without isolated points; it is totally disconnected and not locally compact. These features are central to their investigation by, for example, Cantor, Dedekind and Hausdorff.

4 Theory Reduction, Number Theory and Set Theory

There is a third issue here, which so far I haven't talked much about explicitly. When we claim to have reduced the rational numbers to Peano arithmetic (A_E) by definitional extension, using the trick of Gödel numbering, we have not only 'lifted' the natural numbers to a system of logical formulae, we have also embedded them in a set theoretic hierarchy, a hierarchy of sets. Neither the logical lifting nor the set theoretic embedding conforms to the model of pure theory reduction, and in both cases Gödel numbering stands sideways from the alleged reduction; moreover, the relations between the lifting and the embedding are unclear. Arithmetic is about numbers; logic is about logical formulae (terms, wffs and proofs); but set theory is about sets. When we talk about the rational numbers as a set definable by a wff in Peano arithmetic (A_E), with a place in the arithmetic hierarchy, we invoke set theory but then set it aside. In Enderton's discussion of first-order languages in Chap. 2 of *A Mathematical Introduction to Logic*, he states matter-of-factly, "It is important to notice that our notion of language includes the language of set theory. For it is generally agreed that, by and large, mathematics can be embedded into set theory. By this it is meant that (a) statements in mathematics (like the fundamental theorem of calculus) can be expressed in the language of set theory; and (b) the theorems of mathematics follow logically from the axioms of set theory" (Enderton 1972: 69). This claim might lead us to believe that arithmetic can be expressed straightforwardly in the language of set theory (via a dictionary), and that its theorems, suitably translated, will follow from the axioms of set theory. But neither of these claims is correct; there are relations of reduction but they are all problematic (too strong and too weak) and so useful for the growth of knowledge.

The arithmetical hierarchy as we have been discussing it is at once a hierarchy in mathematical logic, which sorts well formed formulae according to their internal complexity, and suggests special procedures for reducing the complexity of those objects in order to exhibit their canonical form; and it is a hierarchy of sets of natural numbers, the 'intended model' of the first order theory A_E. Note first of all that this hierarchy is much vaster than schoolroom arithmetic, which takes place mostly at the levels of recursive and recursively enumerable sets; the hierarchy contains an infinite number of levels above those two, which from the point of view of arithmetic are mostly vacuous. Secondly, note that the sets in the arithmetical hierarchy so understood have *Urelemente*, the natural numbers, and therefore presuppose the existence and availability of the natural numbers. But when Enderton, for example, claims that arithmetic can be embedded into set theory, he means that the subject matter of natural numbers can be erased and folded into the subject matter of set theory by means of a suitable dictionary. But then we cannot presuppose the natural numbers as *Urelemente*; the set theoretic hierarchy must be constructed out of the empty set, the set of the empty set, and further iterations created by the operation of set formation, and then correlated with the natural numbers.

When we embed the natural numbers themselves in the set theoretical hierarchy, we get back only an equivalence class of sets, different threads that go up the levels (all of them starting at the empty set Ø) and imitate the natural numbers adequately. There is a variety of possible bridge laws that map the natural numbers in appropriate ways to a specific sequence of sets. Thus for example, the bridge laws might identify 0 with the empty set Ø, 1 with the set of the empty set, 2 with the set of the set of the empty set, and so forth, while working out mappings between the natural numbers along with the operations +, ·, and E, and Ø, {Ø}, {{Ø}}, along with negation, union and intersection, which preserve the laws of arithmetic. But instead of choosing Ø, {Ø}, {{Ø}}, {{{Ø}}},..., we could choose Ø, {Ø}, {Ø,{Ø}}, {Ø, {Ø}, {Ø}}},...., and that would work just as well. Thus the analogue of the natural numbers once number theory has been 'reduced' by set theory is really an equivalence class of sequences of sets, and the criteria for membership in that equivalence class is available not within set theory but only by appeal to arithmetic. We have to know what the natural numbers are to know which of the many sequences of sets are candidates for mimicking them.

The alleged reduction of number theory to set theory can be understood in two different ways. If we think of the reduction in more discursive terms (re-writing discourse about numbers as discourse about sets), it looks like a project undertaken by model theorists to reduce the ontological extravagance of number theory. Then the aim of the reduction is to match the things, operations, and procedures of number theory with counterparts at determinate levels of the set theoretic hierarchy, and further to re-write a given episode in number theory so that the pertinent items correspond to even lower levels, if possible. And the theory reduction is not pure. If we think of the reduction in more ontological terms, however, in the hopes of achieving a pure theory reduction where numbers disappear, it looks like a project undertaken by set theorists whose main concern is not logic, but the desire to find a unified theory for all of mathematics. In this case, we are reminded that the area of mathematics with the most extravagant ontology of all is set theory (with the possible exception of category theory), because the mysterious operation of 'power set' allows the mathematician to go ever up and up the hierarchy of transfinite cardinals. This is of course why set theory can hope to include an analogue of every other mathematical item. If set theorists wanted to argue that mathematics is really only about sets and thus to offer a unified theory in the language of set theory formulated in terms of predicate logic, it would have to be couched in not just an n-order logic but eventually in a $\aleph_n$-order logic. It is hard to imagine what such a theory would look like on the page. Also, sets of very high cardinality seem to have strange properties (insofar as we can understand them), so the set theorist seems to be trading in one kind of heterogeneity for another; we don't really know what axioms would cover all of set theory. But all along we have been making the assumption that definitional extension will take place at relatively low levels (the lowest of the arithmetic levels) of the set theoretic hierarchy, and can be captured by first-order logic, since it is really only within first-order formal languages that we know precisely what we mean by 'successful deduction.'

There are two issues here. We must distinguish between the logical complexity of the object, and the logical complexity of reasoning about the object. For example, the real numbers are the first and most important example of a set that is demonstrably not countable; it does not exist anywhere in the arithmetical hierarchy. We must go up to the analytical hierarchy, where sets of natural numbers are defined by wffs from second-order predicate logic, in order to locate the real numbers. However, if we accept the real numbers (despite their logical complexity), we find that the first-order theory of real closed fields has very nice properties. Any real closed field F has the same first-order properties as the field of real numbers; examples include the field of real algebraic numbers, and the field of the non-standard reals. Tarski proved that the theory of real closed fields in the first order language of partially ordered rings (consisting of the binary predicate symbols = and $\leq$, the operations of addition, subtraction, multiplication and the constant symbols 0 and 1) admits elimination of quantifiers. The important consequences of this result for model theory is that the theory of real closed fields is complete, o-minimal (o-minimality is a weak form of quantifier elimination) and decidable, that is, there is a well-defined algorithm for determining whether a sentence in the first-order language of real closed fields is true (see Tarski 1951; Sinaceur 1999: Part IV, Chap. 1). Euclidean geometry, without the ability to measure angles, is also a model of the real closed field axioms, and so is also decidable. Model theorists are often happy to put the issue of the logical complexity of the object in brackets, and to look at the logical complexity of reasoning about it; as interesting and productive as this approach is, the 'reduction' involved doesn't meet the standards of pure theory reduction, which would like the real numbers to be folded back into the rationals, thence to the natural numbers, thence to sets. The other option is to look for an intuitionist or constructivist analogue of the real numbers (and of the procedures of infinitesimal analysis), which might plausibly live in the arithmetical hierarchy, and try to enforce a restriction of mathematical activity to them; the difficulty is then that practicing mathematicians, while they are sometimes interested in knowing the level of logical complexity of their objects and procedures, are in general not interested in constructivist restrictions.

Even as astute a philosopher as Butterfield seems to have been persuaded that by turning the natural numbers into sets, by 'covering' the natural numbers with a first-order theory (first order Peano arithmetic) that includes as well the integers and the rational numbers, and by 'covering' in a different sense the real numbers and part of geometry with a well-behaved first-order theory (the theory of closed real fields), that we have mastered most of mathematics by a series of definitional extensions that begins with set theory. However, as Butterfield says about physics, this appearance is an illusion. The results of model theory are striking and important, but they don't mean that geometry, real and complex analysis, and number theory have been reduced to set theory. Rather, these results mean that logicians have a useful way to measure the 'logical complexity' of items in other branches of mathematics, a measurement that can enhance research in a variety of domains, but which may or may not be key or even pertinent to the solution of important problems.

The (illusory) story just told depends on the avoidance of talking about the things of mathematics in favor of talking about discourse about mathematics. It is much easier to believe that discourse about geometry can be reduced to discourse about sets than to believe that figures like circles and triangles can be reduced to sets, especially sets constructed from iterations of bracketing the empty set. Circles and triangles are shapes with distinctive properties *as shapes*: the circle is perfectly and infinitely symmetrical around its center, and the triangle is the most economical way to bound a finite area on the plane with line segments—only three of them (and right triangles are the canonical form of the triangle). You can operate however you want on $\emptyset$, $\{\emptyset\}$, $\{\{\emptyset\}\}$, ... but you will never arrive at a shape.

Let us return to Butterfield's earlier observation about T_1 and T_2, the reduced and the reducing theories, and paraphrase it so that it emphasizes the things the theories are about. Then we conclude, echoing him, that there may well be aspects of geometry, viewed as a problem-solving enterprise that investigates figures, which are crucial to its functioning as a scientific theory, but are not encompassed by (are not part of) the corresponding aspects of set theory. The properties of geometrical figures might simply be different from those of sets; or geometry might have aspects to do with explanation, modelling or heuristics that are not encompassed by or even perceptible in set theory. We cannot 'deduce' apples from oranges; no claims about

Ø, {Ø}, {{Ø}}, …, no matter how elaborated, will yield a claim about the circle. This does not mean, however, as Butterfield also observes, that a theory about geometrical figures and a theory about sets are wholly incommensurable, and that seeking a systematic connection between them is in vain. It may be quite illuminating for a geometer to know that a fragment (and which particular fragment) of a certain formalization of Euclidean geometry is decidable.

Likewise, it is easier to believe that discourse about the natural numbers can be reduced to discourse about sets than to believe that the natural numbers themselves can be reduced to sets. The natural numbers are canonical because they are the only mathematical items that index themselves; they are used to index all other countable things in mathematics, including sets! Thus, in order for iterations of bracketings of the empty set to represent the natural numbers within set theory, we have to be able to count the iterations; the natural numbers are epistemically prior to, and required for, the construction of the sets used to represent them. The natural numbers are prior to any iteration and indeed any hierarchy. Moreover, the natural numbers and the rational numbers lend themselves to, and suggest, embeddings in complex analysis and in algebraic geometry that no thread of iterations of forming sets of the empty set would ever lead to. What a mathematical thing might be, is indicated not only by the solved problems in which it figures but also in the unsolved problems to which it lends itself.

In sum, echoing Butterfield again, there may well be aspects of number theory viewed as a problem-solving enterprise that investigates natural numbers (and by 'impure' definitional extension the integers and the rational numbers), which are crucial to its functioning as a scientific theory, but are not encompassed by (are not part of) the corresponding aspects of set theory. The properties of natural numbers, integers and rational numbers might simply be different from those of sets, especially sets built on the empty set; and problems about them in number theory might have aspects to do with explanation, modelling or heuristics that are not encompassed by or even perceptible in set theory. This does not mean, however, that number theory and a theory about sets are wholly incommensurable, and that seeking a systematic connection between them is in vain. It may be illuminating for a number theorist to know that some of the machinery used in the proof of Fermat's Last Theorem can be replaced by other means of more modest logical complexity.

Part of the genius of logic, is to treat things as if they were discourse. Logic especially encourages this tendency because for valid deductive arguments to go through, the terms must be homogeneous throughout; and things (even mathematical things) tend to be quite heterogeneous. Line segments are not numbers, and numbers are not sets. And subject terms play a role in thought very different from predicates (and relations), a difference which it suits logic to minimize. Discourse is a great homogenizer, and logical discourse all the more so. However, this process of making things discursive not only helps to solve problems, but also generates certain illusions and leads us astray if we are not careful. It is what we do every time we think 'S is P.'

Encountering a problematic thing (which exists in the many ways that things exist, not as isolated but in relation to other things, and offering an irreducible, infinitely complex, unity to our awareness), we look for its conditions of intelligibility. Sometimes, however, having translated the thing upwards into the predicates and relations that articulate its intelligibility, we confuse it with those conditions, or we neglect to notice when we bring it back down that it has become a shadow of its former self or an equivalence class or a useful analogue $S^* \neq S$ of what we started out with. Thus, not only am I arguing that we should keep in mind the distinction between things and discourse, even in mathematics, but also that we might consider problem reduction as even more centrally important for philosophy than theory reduction. Then we will find that the natural numbers, which seem initially so simple and transparent, both inspire and resist the discourses into which we lift or plunge them to solve problems about them: algebraic number theory, analytic number theory, and logic itself. In each case, there are aspects of numbers and the problems they raise that are not encompassed by the 'reducing' discourse, which nonetheless because of the way it represents or highlights other aspects may sometimes offer fruitful problem-solving strategies.

5 Number Theory and Ampliative Explanation

What about the canonical problems of number theory, like those that puzzled Fermat, discussed earlier in this essay? The solution of those problems also offered *explanations* of why those problems arose in the first place, and brought two different ways of looking at the integers into novel relation. In order for these insights to be developed, new concepts had to be articulated, with the creation of novel clusters of concepts. Before we move to the 19th c., however, we should note a few important developments between Fermat and Gauss. In the 17th c., the natural numbers took on a new meaning when they were viewed as solutions to polynomial equations. The advent of polynomial equations made visible the distinctions among rational, algebraic and transcendental numbers, and then led to the dawning realization that many important curves were not associated with any polynomial of fixed degree with a finite number of terms: beyond the algebraic functions lay the transcendental functions. The infinitesimal calculus, and the notation and methods associated with differential equations, allowed for the exploration of these functions by Leibniz, the Bernoullis, Euler and Lagrange in conjunction with the nascent discipline of Newtonian mechanics. This set the stage for Gauss's work, which included the flowering of complex analysis based on the insight that the Euclidean plane provides a good (geometric) model for the complex number system. Gauss also revisited the ancient **Chinese Remainder Theorem** with a fresh conceptualization called the **Quadratic Reciprocity Theorem**, which explained for the first time why congruences, especially congruences precipitated by prime numbers, lie at the heart of number theory. The exploration of congruences reveals the fine structure of the natural numbers.

Albrecht Fröhlich and M.J. Taylor begin their textbook *Algebraic Number Theory* with a nice example of ampliative reasoning, where a problem is carried up into a broader context and then re-situated (Fröhlich and Taylor 1991: 1–2). Exactly the same problem shows up at the beginning of *Number Theory 1* authored by Kazuya Kato, Nobushige Kurokawa, and Takeshi Saito (translated into English by Masato Kuwata) (Kato et al. 1996: 4–5). The problem was first articulated by Fermat in the seventh of the forty-eight comments he left in the margin of his copy of Diophantus' *Arithmetica*, published posthumously by his son. The context for this result was Fermat's study of right triangles in relation to prime numbers. Studying the triplets of integers that satisfy the equation $x^2 + y^2 = z^2$, he noted that the length of the hypotenuse in each case was a prime number congruent to 1 (mod 4) and was never a prime number congruent to 3 (mod 4). To say that two integers are congruent "modulo n" means that if they were divided by n, an integer, the remainder in both cases would be the same; thus the odd primes 5, 13 and 17 (for example) are congruent to, or equivalent to, 1 mod 4. It is easy to see that every integer n thus divides up the infinite set of integers into n equivalence classes, $\mathbf{Z}/n\mathbf{Z}$, which can then themselves be treated as elements of a set; indeed, they form a finite group. (If $n = p$, p prime, then these equivalence classes $\mathbf{Z}/p\mathbf{Z}$ form a finite field.) This sorting of the integers by 'modding out' a subgroup provides useful information, especially when n is prime, or the power of a prime. Note that whereas the concept of a positive integer had an obvious Euclidean-geometrical interpretation (a certain line length, built up from unit line segments), the concepts of a prime number and of congruence did not; here Fermat brings these concepts into a productive association with geometry for the first time.

This result led Fermat to ask, under what condition an odd prime p may be written as the sum of two squares ($p = x^2 + y^2$ where x and y are integers), and under what condition it may not be so written. (Note that this equation has no Euclidean-geometrical significance, though it is clearly formally or rather algebraically related to the equation $p^2 = x^2 + y^2$.) Fermat claimed that the first condition is $p \equiv 1 \pmod 4$ and the second is $p \equiv 3 \pmod 4$ in his Two Squares Theorem. He asserted this claim without proof; Euler proved it about a hundred years later by the formal and rather unilluminating method of infinite descent. But the *key explanatory insight* was that a prime number p congruent to 1(mod 4) loses its irreducibility as a prime number in the ring $\mathbf{Z}[i]$ of Gaussian integers in the field $\mathbf{Q}[i]$, while a prime number p congruent to 3(mod 4) remains prime. That is, the fact that a number is prime is not simply an inherent feature of the number, but also depends upon the number system in which the number is located! This is rather astonishing. Moreover, it accounts for many puzzling features of the integers.

These explanatory insights require some definitions, some of which were introduced in Chap. 1, but which are worth rehearsing here. The field $\mathbf{Q}[i]$ is an algebraic extension of the field of rational numbers $\mathbf{Q}$, obtained by adjoining the square root of -1 ($\sqrt{-1} = i$) to the rationals. Its elements are of the form $a + bi$ where a and b are rational numbers, which can be added, subtracted and multiplied according to the usual rules of arithmetic, augmented by the equation $i^2 = -1$; and a bit of computation shows that every element $a + bi$ is invertible. Within this field $\mathbf{Q}[i]$ we can locate the analogue of the integers $\mathbf{Z}$ within $\mathbf{Q}$: it is $\mathbf{Z}[i]$, the Gaussian integers, whose elements are of the form $a + bi$ where a and b are integers. Like $\mathbf{Z}$,

Z[i] enjoys unique prime decomposition, but its primes are different from those of **Z**; and instead of two units it has four.

To be a prime in **Z**[i], as noted earlier, $a + bi$ must satisfy these conditions: either a is 0 and b is a prime in **Z** and is congruent to 3(mod 4); or b is 0 and a is a prime in **Z** and is congruent to 3(mod 4); or neither a nor b are 0 and $a^2 + b^2$ is a prime in **Z** and is not congruent to 3(mod 4). If we return to Fermat's Two Squares Theorem, we note that 5, 13 and 17 cease to be primes in **Z**[i] and may be factored there, whereas 7, 11 and 19 remain prime.

$$5 = (2 + i)(2 - i) = 2^2 + 1^2$$
$$13 = (3 + 2i)(3 - 2i) = 3^2 + 2^2$$
$$17 = (4 + i)(4 - i) = 4^2 + 1^2$$

Richard Dedekind, using a result of Lagrange, offered a proof that exploits this insight in 1894; here is a version of that presentation is given by Fröhlich and Taylor.

We begin in **Z**. Suppose $p = x^2 + y^2$ (p an odd prime, x and y integers). Consider the sequence of squares: 1, 4, 9, 16, 25, 36, 49, 64, 81, and so forth; it is easy to see that for every even integer n, $n^2 \equiv 0 \pmod 4$ whereas for every odd integer n, $n^2 \equiv 1 \pmod 4$. Moreover, x and y can't both be even, and x and y can't both be odd (because in either of those cases $x^2 + y^2$ would be even, but $x^2 + y^2$ must be equal to an odd prime); so x must be odd and y must be even, or vice versa: in either case $p = x^2 + y^2 \equiv 1 \pmod 4$.

Proving the converse is the hard part. Recall that if we mod out the integers by a prime p, the equivalence classes will form not just a finite group under addition, but also a finite field under addition and multiplication. We assume here that p is an odd prime with $p \equiv 1 \pmod 4$, and we want to show that $p = x^2 + y^2$ with x and y integers. In a finite field F with p elements, the non-zero elements form a multiplicative group of order $(p - 1)$; the order of a group or a subgroup is its cardinality. The order of an element g of a group is the smallest integer n such that $gn = 1$ (the multiplicative identity). Two striking facts about group theory is that the order of every element in the group, as well as the order of every subgroup, divides the order of the group. The order of every element in this group is then a divisor of the integer $(p - 1)$, so that every element satisfies the equation $x^{p-1} = 1$, and is thus a $(p - 1)$th root of unity; so we can find a primitive $(p - 1)$th root of unity, the generator, which makes the group cyclic. (A cyclic group is a finite—or countably infinite group—whose elements are generated by a single element, the generator of the group.) Since $p \equiv 1 \pmod 4$, in this case we can infer that this cyclic group has an element of order 4. The equivalence class that contains -1 is the unique element of order 2 in the finite field of p elements, so we can infer that $m^2 \equiv -1 \bmod (p)$ for some integer m; then p divides $m^2 + 1$.

Now we go upstairs to **Z**[i]. We have just seen that p divides $m^2 + 1$, so it also divides the product of Gaussian integers $(m + i)(m - i)$, a possibility that only appears in the ring **Z**[i]. The proof goes through from this point on because of the happy condition that one can define a Euclidean norm on **Z**[i], $N(z) = z\bar{z}$ or

$N(x + iy) = x^2 + y^2$. (The symbol $\bar{z}$ means the complex conjugate of z.) This makes $\mathbf{Z}[i]$ a principal ideal domain—every element has a unique decomposition into prime elements—and so enjoys a precise analogue of the unique prime factorization found in **Z**. Moreover, z is a unit of $\mathbf{Z}[i]$ if and only if $N(z) = 1$, that is, $z = 1, -1, i$, or $-i$.

Then we proceed by a short *reductio ad absurdum* argument. Suppose p divided either of the factors $(m + i)(m - i)$, then by applying complex conjugation it would divide the other and so p would divide $2i$, which is absurd; so p does not divide either of the factors. Thus p, though it is prime in **Z**, is not a prime element in $\mathbf{Z}[i]$, and there must be a factorization $p = (x + iy)(x' + iy')$ as a product of two Gaussian integers, neither of which is a unit. Indeed, in view of the norm (which is multiplicative, and defined so that $N(p) = p^2$), the factorization can have only two factors and it must be of the form $p = (x + iy)(x - iy)$. Now we go back downstairs. This transit is accomplished by applying the norm to both sides of the factorization of p, to arrive at the equation $p^2 = (x^2 + y^2)(x^2 + y^2)$, and conclude that $p = (x^2 + y^2)$ (Fröhlich and Taylor 1991: 1–2).

The proof strategy sets the concepts involved in this problem of number theory into novel proximity with a variety of concepts. It relates the integers to the complex numbers, since $\mathbf{Q}[i]$ is a subfield of **C**, the field of complex numbers. If we use the Euclidean plane as a model for **C**, the units are then modeled by the square with endpoints $1, i, -1$ and $-i$. This suggests, as we have seen, the generalization that models the set of nth roots of unity as vertices of regular n-polygons centered at 0 on the complex plane, with one vertex at 1. As noted earlier, we can then move from $\mathbf{Q}[i]$ to other algebraic fields called cyclotomic fields, where an nth root of unity (ζ_n) is adjoined to **Q**; for each cyclotomic field $\mathbf{Q}[\zeta_n]$ there is a group of automorphisms that permutes the roots of unity while mapping **Q** to itself, the **Galois group**, Gal $(\mathbf{Q}[\zeta_n]/\mathbf{Q})$. The field $\mathbf{Q}[i]$ is also a vector space, which is not merely an abstract structure, but a toolkit of related concepts, like basis, norm, inner product, vector, scalar, eigenvector and eigenvalue and of course dimension. (The concept of norm plays an important role in the proof just mentioned.) Finally, the notion of congruence plays a central role in the proof, specifically congruence mod 4. We have seen that the key to the latter problem solution is the insight that some primes in **Q** remain prime in $\mathbf{Q}[i]$ and some primes decompose into factors (factors which are prime in $\mathbf{Q}[i]$), and that this sorting must be carried out by looking at the primes in terms of which congruence class (mod 4) they belong to.

As it turns out, the Galois group Gal $(\mathbf{Q}[\zeta_n]/\mathbf{Q})$ is isomorphic to the finite abelian group $(\mathbf{Z}/n\mathbf{Z})^{\times}$, the multiplicative group of $\mathbf{Z}/n\mathbf{Z}$. There also exists a one-one correspondence between subfields of $\mathbf{Q}[\zeta_n]$ and subgroups of Gal $(\mathbf{Q}[\zeta_n]/\mathbf{Q}$. These correspondences allow us to find patterns linking the decomposition (factoring) of prime numbers in cyclotomic fields to specific congruences. For example, a prime decomposes in $\mathbf{Q}[\zeta_7]$ if it is congruent to 1 (mod 7); a prime decomposes in $\mathbf{Q}[\zeta_5]$ if it is congruent to 1 (mod 5). Here is the theorem that generalizes these facts: Let n be a natural number; suppose that a subfield **L** of $\mathbf{Q}[\zeta_n]$ corresponds to a subgroup **H** of $(\mathbf{Z}/n\mathbf{Z})^{\times}$ [isomorphic to Gal $(\mathbf{Q}[\zeta_n]/\mathbf{Q})$]. Then for any prime number p not dividing n, we can prove that p is totally decomposed in **L** if and only if p mod n is an element of **H**. (To say that p is totally decomposed means in this context that it splits into prime factors; in more general contexts, it means that it splits into prime ideals.) (Kato et al. 2010: 16f.) These kinds of patterns, relating 'outer' to 'inner,' that is, the ways that the integers can be embedded in algebraic number fields to the internal fine structure of the integers revealed by congruences, lead eventually to Class Field Theory. As Kazuya Kato writes, rather poetically, in Chap. 8 of his co-authored *Number Theory 2*, "In a fairy tale, sceneries far away from home may be seen in the magic mirror. How many abelian extensions are there over a global or a local field **K**, and what happens in such abelian extensions? These 'outdoor sceneries of **K**' are reflected in the 'indoor mirror of **K**,' that is, the multiplicative group or the idele class group of **K**. This is the main content of what class field theory is" (Kato et al. 2010: 151).

Gauss introduced the Gaussian integers $\mathbf{Z}[i]$ in his monograph on biquadratic or quartic reciprocity (1832), which investigates the solvability of the congruence $x^4 \equiv q \pmod{p}$ with respect to that of the conguence $x^4 \equiv p \pmod{q}$ (where p and q are distinct odd primes greater than two). He found that solving this problem was easier if he first posed it in terms of the Gaussian integers rather than ordinary integers. This work followed his investigations into quadratic and cubic reciprocity, analogously stated for x^2 and x^3. As it turns out, every quadratic field is a subfield of a certain cyclotomic field (this is an instance of the **Kronecker-Weber Theorem**); thus the way that a prime number p decomposes in a quadratic field is also determined by p mod n for some n, and the Quadratic Reciprocity Theorem may be understood as a re-statement of this fact. Class field theory generalizes this result, to cases where the ring of integers in an algebraic field does not enjoy prime decomposition, but only decomposition into **prime ideals**.

Fröhlich and Taylor give the following account of this generalization (here **K** stands for any algebraic number field and $\mathbf{O}_K$ for its ring of integers): "Whilst rings of algebraic integers are not in general principal ideal domains, and so do not possess unique factorisation of elements, they do still possess unique factorisation of non-zero ideals: that is to say, given a number field **K**, every non-zero $\mathbf{O}_K$-ideal can be written uniquely (up to order) as a product of prime ideals of $\mathbf{O}_K$." (Fröhlich and Taylor 1991: 4). The extent to which unique factorization fails in the ring of integers of an algebraic number field can be described by a certain group known as the **ideal class group**; this group is finite, and its order is the class number, which registers the extent to which prime factorization fails. (Wiles' final struggle to show that T_Σ and R_Σ are isomorphic, noted in Chap. 3, involved the creation of a class number formula.) The class group of $\mathbf{O}_K$ is defined to be the group of **fractional** $\mathbf{O}_K$-ideals modulo the subgroup of **principal** fractional $\mathbf{O}_K$-ideals, where **K** is an algebraic number field. When every $\mathbf{O}_K$-ideal is a principal ideal, or when we can define a Euclidean norm on $\mathbf{O}_K$, this class group is trivial, that is, it is just the (multiplicative) group consisting of the element 1, and thus the class number is one and we have unique prime decomposition. In sum, the task of locating and defining structure in rings of algebraic integers within algebraic extension fields that will be analogous to structure in the ordinary integers turned out to be very difficult. My point is, once again, that going upstairs from **Z** or **Q** to algebraic number fields like **Z**[i] within **C** in order to solve certain problems that arise early on in number theory (already in the notes of Fermat) is a non-trivial matter: the reasoning is ampliative, both going up and coming down.

Chapter 4 of Fröhlich and Taylor is devoted mostly to units and class groups, though with an oddly geometrical section on lattices in Euclidean space, noteworthy because Fröhlich and Taylor try very hard to keep geometry and real and complex analysis out of this book: this is a book about *algebra.* They then conduct the reader into a more detailed study of fields of low degree (Chap. 5 deals with quadratic, biquadratic, cubic and sextic fields), and then of cyclotomic fields (Chap. 6), where, significantly, quadratic fields are re-visited and the quadratic reciprocity law is re-proved. It was Gauss's favorite theorem; in the *Disquisitiones Arithmeticae* (1801) he called it the fundamental theorem and by the time he died he had proved it eight different ways. The next step up in complexity—and it is a considerable step—is to cubic fields, created by adjoining to **Q** the roots of equations whose exponents may go up to 3. Recall that the order of an element a in a multiplicative group G is the minimum positive integer m such that $a^m = e$, the identity element. In general, the group of roots of unity in a field **L** is a subgroup of the points of finite order of the multiplicative group $\mathbf{L}^{\times}$. If we adjoin a primitive nth root of unity ζ_n to **Q**, for example, we adjoin the whole group of nth roots of unity, because ζ_n generates the whole group. Analogously, we can explore what happens when one adjoins the coordinates of the group G of points of finite order (torsion points) of an elliptic curve to a field **L**; this generates a theory analogous to the theory of cyclotomic fields. Cyclotomic fields, as we have seen, have very nice properties. The ring of integers in the cyclotomic field $\mathbf{Q}[\zeta_n]$ is just $\mathbf{Z}[\zeta_n]$ and the abelian Galois group Gal $(\mathbf{Q}[\zeta_n]/\mathbf{Q})$ is isomorphic to the multiplicative group $(\mathbf{Z}/n\mathbf{Z})^{\times}$, as we noted earlier. The Kronecker-Weber Theorem tells us that every Abelian extension of **Q** (and thus every quadratic extension of **Q**) is contained in some cyclotomic field $\mathbf{Q}[\zeta_n]$. Because of these nice properties, cyclotomic fields can be described and classified in terms of objects defined back down in **Q**, residue class characters, elements of the 'character group' associated with $(\mathbf{Z}/n\mathbf{Z})^{\times}$, another version of the way that 'inner' is related to 'outer' in algebraic number theory.

6 Coda

Peter Roquette makes a rather Spinozan observation at the beginning of his interesting long essay, "What is Reciprocity? On the Evolution of class field theory in the 20th century," when he remarks: "… the mere proof of the validity of a theorem is in general not satisfactory to mathematicians. We also want to know "why" the theorem is true; we strive to gain a better understanding of the situation than was possible for previous generations. Sometimes a result seems to be better understood if it is generalized, or if it is looked at from a different point of view, or if it is embedded into a general theory which opens analogies to other fields of mathematics". (This essay is under revision and no longer available online. Please see: http://www.rzuser.uni-heidelberg.de/~ci3/manu.html.) And Kazuya Kato sounds quite Leibnizian at the beginning of his co-authored *Number Theory 1*, when he writes, "Fermat, who was the founder of modern number theory, noticed the depth of the world of numbers… We think that the reason for the depth of the world of numbers that fascinated Pythagoras, Fermat, and many others is that it is a reflection of the depth of the universe. As number theory has been developed during the 350 years since Fermat's era, we have discovered the enormous depth of the world of numbers." (Kato et al. 1996: 14).

Interesting mathematical systems are emphatically non-trivial and require first principles whose truth is evident only against a background of highly developed knowledge. And that dependence on background knowledge leaves them open to questioning, a feature that enhances rather than impairs their status as axioms. Good axioms (take the Peano Postulates, for instance!) summarize and refer back to a great deal of hard-won mathematical experience, and that very reference can be used to adjust, unpack, generalize and improve them. But no matter which way you look at it, a number is not a set and it is not a shape; however, numbers acquire new meaning by being brought into relation with things that are not numbers, as when the integers are embedded in various algebraic number fields as well as in the complex plane. Look as long as you want at the Peano Postulates, or at the hierarchy of sets built on the empty set: you will not find a square or a circle there. (Nor, for that matter, will you find a vector space, a norm, a Galois group, a reciprocity theorem, or the complex plane.) The Peano Postulates put important constraints on what might count as the integers, and perhaps one of the structures of this theory is classroom arithmetic; but they do not give us number theory. And we should keep in mind that even classroom arithmetic, the kind we offer to our children in primary and secondary school, is riddled with inexplicable mysteries. Why can we easily find integral solutions for $x^2 - y^2 = z^2$ but none at all for $x^3 - y^3 = z^3$? Why can some primes be written as the sum of two squares x and y (where x and y are integers), but other primes can't? If we were limited to the language (and concepts) of Peano Arithmetic, we would never be able to explain. Problem solving and explanation in mathematics are inherently ampliative, and that is what makes mathematical research so much more exciting than what we do in the schoolroom, deducing theorems from neatly arranged axioms: satisfying as that activity is, deduction should ultimately inspire students, and all lovers of mathematics, to go beyond deduction.

References

Butterfield, J., & Isham, C. J. (1998/1999). On the emergence of time in quantum gravity. In J. Butterfield (Ed.), arXiv: gr-qc/9901024v1. Repr. in *The Arguments of Time* (pp. 111–168), Oxford: Oxford University Press).

Enderton, H. B. (1972). *A mathematical introduction to logic*. New York: Academic Press.

Feyerabend, P. (1962/1981). Explanation, reduction and empiricism. In H. Feigl & G. Maxwell (Eds.), *Minnesota Studies in Philosophy of Science* (Vol. 3). Minneapolis: University of

Minnesota Press, 1962: 28–97. Repr. in Feyerabend, F., *Realism, rationalism and scientific method*. Cambridge: Cambridge University Press.
Fröhlich, A., & Taylor, M. J. (1991). *Algebraic number theory*. Cambridge: Cambridge University Press.
Grosholz, E. (1991). *Cartesian method and the problem of reduction*. Oxford: Oxford University Press.
Hicks, A. (2013). *Ampliative inference and the principle of sufficient reason*. Ph.D. thesis, State University of New York at Buffalo. ProQuest 3613007.
Kato, K., Kurokawa, N., & Saito, T. (1996). *Number theory 1: Fermat's dream*. M. Kuwata and K. Nomizu (Tr.). Providence, R. I.: American Mathematical Society.
Kato, K., Kurokawa, N., & Saito, T. (2010). *Number theory 2: Introduction to class field theory*. M. Kuwata & K. Nomizu (Tr.) Providence, R. I.: American Mathematical Society.
Nagel, E., & Newman, J. (1958). *Gödel's proof*. New York: New York University Press.
Nagel, E. (1961). *The structure of science: Problems in the logic of scientific explanation*. New York: Harcourt, Brace & World.
Sinaceur, H. (1999). *Corps et modèles*. Paris: Vrin.
Tarski, A. (1951). *A decision method for elementary algebra and geometry*. Berkeley: University of California Press.

Chapter 5
Fermat's Last Theorem and the Logicians

1 Logic and the Illusion of Homogeneity

At the risk of trying the reader's patience, in this chapter I am going to revisit *A Mathematical Introduction to Logic* (Enderton 1972), and *Gödel's Proof* (Nagel and Newman 1958), before examining the first stage of Wiles' proof of Fermat's Last Theorem as a case study. I return to Butterfield's critique of the mid-twentieth century model of theory reduction, but whereas in the last chapter I tried to show the limitations of the theory reductions presented by Enderton and in a different sense by Gödel in Butterfield's own terms ("too weak and too strong"), here I want to show how these reductions (imperfect as they are) lend themselves to the growth of knowledge, often by means of the productive ambiguity that occurs when disparate discourses must meet and mingle. And this sheds light in retrospect on the episode from number theory presented at the very end of Chap. 4, which sets the stage for my exposition of the first stage of the proof of Fermat's Last Theorem.

Logic, in order to formulate rules of correct inference, must abstract from all subject matter. In order to offer rules of thought in general, it must not express special assumptions about special features of this or that kind of thing. The AAA1 syllogism reliably transmits truth (when the premises are true) no matter what kinds of terms are substituted for S, M, and P; modus ponens reliably transmits truth (when the premises are true) no matter what kinds of propositions are substituted for p and q. Predicate logic seems to introduce considerations of subject matter because of the quantifiers, which must quantify over a universe of discourse, but students introduced to 'pure predicate logic' in a textbook are told to assume that the universe of discourse includes everything. Robert Paul Churchill, in his standard textbook, *Logic. An Introduction* (1990), for example, advises the student that we must establish a 'universe of discourse' which includes everything that can be named or described, whether physical or non-physical (Churchill 1990, 149). By adding special vocabulary to the language of pure predicate logic, we may develop

E.R. Grosholz, *Starry Reckoning: Reference and Analysis in Mathematics and Cosmology*, Studies in Applied Philosophy, Epistemology and Rational Ethics 30, DOI 10.1007/978-3-319-46690-3_5

a formal language that restricts predicate logic to a certain subject matter (arithmetic, for example); in textbook problems, restrictions are introduced informally by telling students that the universe of discourse has been chosen to include only certain kinds of things (all human beings, for example).

This fact about logic has led some philosophers to claim that logic is purely syntactic, and that the regulation and organization it provides for human knowledge operates only at the level of syntax. The theory of the syllogism, propositional logic and pure predicate logic are intended to introduce no restrictions on subject matter, but to remain semantically neutral (Carnap 1928, 1983). Other philosophers like Bas van Fraassen, wishing to introduce semantical considerations, presented models for logical theories that are themselves constituted in terms of formal languages, so that the relation between model and formalized theory depends on an isomorphism between object language and meta-language. (Van Fraassen 1989: Part III; he thinks beyond the 'myth of pure syntax' in his later book, Van Fraassen 2008.) Because first order theories have nicer properties (and so are better understood) than second order theories, the preferred formal theories are usually expressed in first order pure predicate logic augmented by special symbols, along with rules for the employment of the special symbols. However, when they are intended to be about mathematical systems, these first-order theories are non-categorical, that is, they are satisfied by an infinite spectrum of isomorphic models. The models, even the somewhat mysterious 'intended model,' thus also come to seem like uninterpreted structures, syntactic objects. And structuralism has come to seem like an attractive option for philosophers of mathematics.

However, let us step back from the myth of pure syntax, and from the assumption that logic is the main—perhaps the sole—source of organization and regulation for human (and mathematical) thought. Then we may see that the central task of syllogistic, propositional logic and pure predicate logic, which is to give rules for the deductive transmission of truth, itself imposes restrictions on subject matter, and on the relations between a formal theory and subject matter. The first is the restriction of homogeneity on the subject matter. Logic, to exhibit the forms of deductive inference and to serve as the standard of argument, must be expressed in a homogeneous idiom: not only must p, q, and r be homogeneous, but so must be S, P and M. Since they are the same kind of thing, S and P can be yoked unproblematically in the universal proposition, 'All S is P,' and S, P and M can be used either as the subject or the predicate term in a proposition. Likewise, p, q, and r can be inserted within a premise or a conclusion as needed. In predicate logic, constant terms and n-place predicate terms are distinguished, but their homogeneity is guaranteed because, when and if interpreted, they are referred to a single universe of discourse whose elements must all be logically compatible. The extensional interpretation of predicate logic guarantees that constant terms and n-place predicate terms are all taken to be elements and subsets of the same universe of discourse, and thus for example a constant term a either will or will not occur in a one-place predicate term H. If it does, then Ha is true; if it doesn't, then Ha is false.

The second restriction is that the relations between logical items and subject matter must be univocal and stable: they must be expressible in terms of an isomorphism. Whenever S or P occur, or *p*, *q*, and *r*, they must be exactly the same item in the formal language; likewise, once the issue of subject matter has been reinstated, they must always be interpreted in only one way, and indeed in exactly the same way at every occurrence. These two restrictions hold because otherwise logic cannot do its job: they are required in order for a deductive argument form to transmit truth reliably. If the homogeneity condition did not hold, we could not guard against the fallacy of irrelevance, beloved of religious fundamentalists; if the univocality condition did not hold, we could not guard against the fallacy of equivocation, beloved of politicians. We must be able to combine and transpose and re-identify the items of our logic as we reason.

During the last century, these formal requirements of logic, which it must invoke in order to be logic, acquired a life of their own and turned into formal conditions that bore upon subject matters. This happened both because predicate logic became central to philosophy of science and mathematics, and because logical positivists tended to re-fashion subject matters to look like formal theories. We can see the demands for homogeneity and univocality operating on many different levels. It produced the mid-century model of theory reduction; the deductive-nomological model of explanation and prediction; and an extensional, set-theoretical understanding of predication. But this widespread tendency to impose logical restrictions on subject matters has made many important aspects of scientific and mathematical rationality hard to see, or has erased them altogether. One aspect, diagnosed very well by Robin Hendry, is that scientific and mathematical definition and explanation often exhibit asymmetries that are central to the reasoning, but which are erased by the symmetry of 'isomorphism' when it is used to explicate them philosophically (Hendry 2001). Another aspect, which I have tried to show in the preceding chapters, is that scientific and mathematical activity often make use of subject matters and associated discourses that are logically (though not, of course, rationally) disparate, and this occurs at the level of theory, argument, and proposition.

In such theory integration and in such explanations, terms may fail to be univocal; I have also studied the uses of what I call productive ambiguity in various proofs, where terms are required to shift their meanings in order for the proof to go through (Grosholz 2007: Chaps. 9 and 10). But the interesting question is how these shifts take place, and how they can deepen rather than subvert understanding. I have been arguing that the study of these compelling and up till now often invisible questions should be a matter not only of philosophical but also of historical interest: we need historical epistemology. Logical positivists often assume that a philosopher who strays from the straight and narrow path of logic must then fall into the swamps of naturalism, the empirical study of human behavior or human perception. But the cure for logicism is not 'naturalism' or 'empiricism,' but rather reflective history. The discipline of history is empirical, as I argued in Chap. 2, but its methodology is distinct from that of the sciences, because it is concerned with human actions and practices that are temporal and cultural; and it is history that philosophy needs as an ally in this project.

2 Enderton Redux

In his well-known logic textbook cited in Chap. 4, *A Mathematical Introduction to Logic*, Herbert B. Enderton introduces the study of models in the following way. He states that a structure for a first-order language will tell us what collection of things the universal quantifier symbol refers to, and what the other parameters (the predicate and function symbols) denote. Then he formally defines a structure U for a given first-order language as a function whose domain is the set of parameters (Enderton 1972: 79).

1. U assigns to the universal quantifier symbol a non-empty set $|\mathrm{U}|$, called the universe of U.
2. U assigns to each *n*-place predicate symbol P an *n*-ary relation P^{U} which is a subset of the set of all *n*-tuples of members of $|\mathrm{U}|$, $|\mathrm{U}|^{\mathrm{n}}$.
3. U assigns to each constant symbol *c* a member $c^{\mathrm{U}}$of the universe $|\mathrm{U}|$.
4. U assigns to each n-place function symbol f an *n*-ary operation f^{U} on $|\mathrm{U}|$, so that f^{U}: $|\mathrm{U}|^{n} \rightarrow |\mathrm{U}|$.

Note that the structure U is a function; in so far as it is a function, it is just as 'discursive' or 'logical' as the so far uninterpreted first order language that serves as its domain. Considered simply as a function which performs a service for predicate logic, the structure U is itself presumably also uninterpreted, even if its service is to provide an interpretation. That is, the structure U considered simply as a function has no ontological import, no more ontological import than the uninterpreted first order language that serves as its domain.

If we then think of the structure U as 'purely' discursive and the set $|\mathrm{U}|$ as somehow outside of or beyond discourse, we must wonder how they have been brought into relation. One answer might be that by treating whatever it is that lies outside or beyond discourse as a set, we have assimilated it to a discourse (the discourse of set theory). But then we have only established a mapping between two discourses: does this mapping really count as successful reference and denotation? Doesn't it seem odd that the discourse of first order predicate logic and the discourse of set theory should resemble each other so closely? Perhaps there is an historical explanation for that notable resemblance.

Another answer might be that we find here an example of two different functions of language, the function of indicating what we are talking about and the function of analyzing it. Then the structure U considered as a function has the job of analysis, and the set $|\mathrm{U}|$ has the job of indicating, or perhaps exhibiting, what we are talking about. But what is the relation between the structure U (which is a function) and the non-empty set $|\mathrm{U}|$, which is presumably more like a referent or object, since it provides the interpretation? This is a very difficult question to answer; in a sense, the philosophically vexed question of how applied mathematics is possible at all is implicated in this question, and our choice of reply to it. What is Enderton's response to it? After a few more pages of exposition, Enderton gives an example of what it means for a model to satisfy a set of sentences.

Example. Assume that our language has the parameters 'universal quantifier,' P (a two-place predicate symbol), *f* (a one-place function symbol), and *c* (a constant symbol). Let U be the structure for this language defined as follows:

$|U|$ = N, the set of all natural numbers;
P^U = the set of pairs of natural numbers <*m*, *n*> such that *m* is less than or equal to *n*;
f^U = the successor function S; and $c^U = 0$.

Then Enderton adds, off-handedly, "We can summarize this in one line, by suppressing the fact that U is really a function and merely listing its components: U = (N, $\leq$, S, 0)." (Enderton 1972: 85).

And thereafter, the fact that U is 'really' a function is always suppressed and Enderton writes as if the structure U can be thought of as the referent, the object of discourse, in this case the natural numbers. So at the beginning of Chap. 3, on Undecidability, Enderton introduces the structure (N, 0, S, <, +, ·, E) as the 'intended structure' for the first-order language of number theory with equality and the usual parameters (Enderton 1972: 174). Clearly here (in an exposition of the undecidability of certain logical theories) the structure U is meant to stand for the referent or object that supplies the interpretation and lies somehow beyond or outside of the discourse of first order predicate logic.

Shall we accuse Enderton of intellectual dishonesty? In fact, what he has done here is to employ a common strategy of mathematicians (and scientists) who must bring an analytic discourse into rational relation with what it refers to. Of course, there are no such things as bare facts or raw data! What is referred to, is never encountered wholly outside of discourse: we articulate our awareness of things in one way or another, according to one or another mode of representation, and thereby develop modes of representation and formal idioms that lend themselves well to indicating what we are talking about. Meanwhile we develop other, disparate modes of representation and formal idioms that lend themselves better to certain kinds of analysis of the things under investigation. Some terms occur in more than one of the disparate discourses that develop, and so may serve as bridges between them, though usually because the mathematicians or scientists learn to live with their ambiguous meaning: they typically mean one thing in one discourse and something slightly different in another. (Thus H_2O means a certain molecule in the middle of a chemical article, and a purified substance in a beaker in the account of the experiment at the end of the article.) Enderton needs the student to believe in Chap. 2 that the structure U is part of the formal logical apparatus, and in Chap. 3 that the structure U is the object of knowledge. His sleight-of-hand is so brief and casual that for most students it goes unnoticed.

The problem with Enderton's exposition is that he has left out the discourses that mathematicians historically employ for referring to numbers. In one sense it is the arithmetic that children learn, expressed in the idiom of the Indo-Arabic numerals enhanced by the polynomials of Descartes and Fermat, which inter alia allows us to express the prime decomposition of a large number in the following perspicuous way: $243{,}000{,}000 = 2^6 \times 3^5 \times 5^6$. In another sense, it is the multiply idiomatic discourse to be found in an article published in a journal devoted to number theory, where mathematicians present and defend their latest results. Recall that for Enderton, the 'intended structure' U = (N, 0, S, <, +, ·, E) has already been assimilated to a logician's discourse in which the numbers are represented as the initial element 0 and successive iterations of the function successor, S. This mode of representation was developed by logicians to subject arithmetic to logical analysis, but it is never employed in the classroom to teach arithmetic, or in articles in journals devoted to number theory. When logicians need to use the natural numbers as indices, or to compute number theoretical facts like the prime decomposition of a large number, they make use of Indo-Arabic/Cartesian notation without really mentioning their own departure from the formalism they are supposed to be using.

This doesn't mean that there is a single preferred idiom for referring or picking out the things we are interested in 'correctly.' Successful referring typically depends on the context of use, and changes in the historical context of problem-solving may lead us to change the representations we use not only for analyzing but also for referring. However, the notation 0, S0, SS0, SSS0, … however useful for logical analysis of the natural numbers, is not useful for successful reference in most problem-solving situations in classroom arithmetic or in number theory.

3 Gödel Redux

An accurate and well-received exposition (cited in Chap. 4) of Gödel's two Incompleteness Theorems is given by Ernest Nagel and James R. Newman in their book *Gödel's Proof*, dedicated to Bertrand Russell (Nagel and Newman 1958: 78–79). The proof strategy is fairly well known. Gödel begins with a theory (a set of axioms and their deductive consequences) in the language of first order predicate logic with parameters like those offered by Enderton. Then he assigns a natural number, now called its Gödel number, to every well formed formula (wff) in such a way that if we are given the number, its prime decomposition will allow us to recover the wff; and then he performs a similar feat for any sequence of wffs. Then he devises a numerical function 'Dem.'

Here is Nagel and Newman's exposition of Dem: "Let us fix attention on the meta-mathematical statement: "The sequence of formulas with Gödel number x is a proof of the formula with Gödel number z." This statement is represented (mirrored) by a definite formula in the arithmetical calculus which expresses a purely arithmetical relation between x and z… We write this relation between x and z as the formula 'Dem (x, z)' to remind ourselves of the meta-mathematical statement to which it corresponds (i.e. of the meta-mathematical statement 'the sequence of formulas [wffs] with Gödel number x is a proof (or a demonstration) of the formula with the Gödel number z')." (Nagel and Newman 1958: 85–97).

This relation Dem (x, y) is used to create the celebrated Gödel Sentence G via a carefully constructed, self-referential designation '(sub(m, 13, m))' which picks out "the Gödel number of the formula that is obtained from the formula with Gödel number *m*, by substituting for the variable with Gödel number 13 [*y*] the numeral for *m*," which in turn can be shown to be a definite number that is a certain arithmetical function of the numbers *m* and 13, and the function itself can be expressed within the formalized system. The Gödel Sentence *G* is then constructed by beginning with the following formula:

$$(x) \sim \text{Dem}\,(x, \text{sub}(y, 13, y)),$$

which meta-mathematically claims that "the formula with Gödel number sub(*y*,13,*y*) is not demonstrable." It has a Gödel number, which we will call *n*. Then the Gödel Sentence *G* is:

$$(x) \sim \text{Dem}\,(x, \text{sub}(n, 13, n)),$$

and the meta-mathematical meaning of *G* is "The formula with Gödel number sub (*n*,13,*n*) is not demonstrable." Gödel has cleverly set up the 'diagonalization,' so that in fact the Gödel number of *G* is sub(*n*,13,*n*), and the formula claims of itself that it is not demonstrable. Gödel then uses this special formula to show that *G* is demonstrable if and only if $\sim G$ is demonstrable and thus that if the axioms of this formalized system of arithmetic are consistent, then *G* must be formally undecidable. And *G* can be shown to be true by meta-mathematical reasoning: it formulates a complex numerical property that must hold of all numbers. So the claim that 'if arithmetic is consistent, it is incomplete,' is represented by a demonstrable formula within formalized arithmetic (Gödel 1931, 1992).

The use of Gödel numbering as the strategic bridge between formalized arithmetic qua logical system, and arithmetic, forces an ambiguity on the natural numbers similar to the ambiguity we noted in Enderton's treatment of the structure U. Enderton uses the notation U = (N, $\leq$, S, 0) to mean both an uninterpreted structure, and to mean the natural numbers, the referents which various formalizations are trying to represent. When he wants to emphasize the 'analytic' meaning, he invokes the notation 0, S0, SS0, etc., and when he wants to emphasis the referential meaning, he invokes the notation 1, 2, 3, etc. Gödel likewise constantly invokes the natural numbers (he was a Platonist, after all), but uses a variety of notations to represent them, depending on whether he is using them in an analytic or referential capacity. This is noteworthy, since he is intent on expressing his proof in the strict notation of *Principia Mathematica* as the title of his proof announces.

Gödel uses a notation slightly different from that used by Nagel and Newman (who have simplified his exposition a bit to make it clearer); I will now refer to the notation in *On Formally Undecidable Propositions in Principia Mathematica and Related Systems*. The natural numbers are represented as $0, f\,0, ff\,0, fff\,0$, etc.; 0 and *f* are included among the 'basic signs', and the series just given is included among 'signs of first type'. Along with the '*n*-type signs,' they constitute the class of elementary formulae; Gödel next gives a set of axioms numbered I–V, and defines the class of provable formulae in terms of them. Then he announces Gödel numbering (Gödel 1992: 41–46).

The basic signs of the system P are now ordered in one-to-one correspondence with natural numbers, as follows:

"0" … 1
"f" … 3
"∼" … 5
"v" … 7
"Π" … 9
"(" … 11
")" … 13

Furthermore, variables of type n are given numbers of the form p^n (where p is a prime number greater than 13). Hence, to every finite series of basic signs (and so also to every formula) there corresponds, one to one, a finite series of natural numbers. These finite series of natural numbers we now map (again in one-to-one correspondence) on to natural numbers, by letting the number $2^{n1} \cdot 3^{n2} \ldots \cdot p_k^{nk}$ correspond to the series $n_1, n_2, \ldots n_k$, where p_k denotes the kth prime number in order of magnitude. A natural number is therefore assigned in one-to-one correspondence, not only to every basic sign, but also to every finite series of signs.

This passage rewards study. When Gödel wants to refer directly to the natural numbers, as in the array where numbers are assigned to 'basic signs,' and whenever he indexes anything, he uses Indo-Arabic/Cartesian notation. He has to do this, since, for example, "0" is assigned "f 0" in the notation of his formal system, which, if assignment is taken to be identification, is the contradiction 0 = 1. He doesn't intend to propose a contradiction, of course, but rather an isomorphism between formulae and natural numbers. All the same, we might think that Gödel does in fact understand this assignment as identification, since a few pages earlier he writes: "Proofs, from the formal standpoint, are likewise nothing but finite series of formulae (with certain specifiable consequences). For metamathematical purposes it is naturally immaterial what objects are taken as basic signs, and we propose to use natural numbers[7] for them. Accordingly, then, a formula is a finite series of natural numbers[8], and a particular proof-schema is a finite series of finite series of natural numbers." (Gödel 1992: 38–39).

He nuances this passage with two footnotes that adumbrate 'natural numbers.' Footnote 7 explains that we are to understand the identification of natural numbers with basic signs as a one to one mapping, despite the clear suggestion that the objects in this proof *are* natural numbers. Footnote 8 is a statement of Gödel's Platonist faith; the natural numbers do not exist in space and cannot be set in a spatial array.

My point is not to accuse Gödel of confusion, or illogic, or metaphysical commitments. Rather, I am trying to show that to carry out his proof, he must use modes of representation that lend themselves to logical analysis (Russell's notation) but not to computating or referring, and other modes of representation that lend themselves to successful reference (Indo-Arabic/Cartesian notation). He must use disparate registers of the formal languages available to him, combine them, and exploit their ambiguity, as he does in the passage just quoted. When Gödel says 'natural number,' he must mean both a series of formulae in the system PM, and the objects of arithmetic, in the language of arithmetic (see also Wagner 2008).

Thus if a Gödel number is to function as a plank in the bridge between the formalized, uninterpreted calculus and arithmetic, Gödel (as well as Nagel and Newman) must invoke it in two ways. It must be a number that we can refer to by a numeral in Indo-Arabic/Cartesian notation, because only in that notation can the wffs be retrieved from their Gödel numbers, via unique prime decomposition. So for example, (here I revert to the exposition of Nagel and Newman where the Gödel numbering is slightly different) the PM formula '0 = 0' in the uninterpreted calculus is represented by $2^6 \times 3^5 \times 5^6$ which is equal to 243,000,000; we must use the prime decomposition in Indo-Arabic/Cartesian notation to rewrite 243,000,000 in order to find the powers of 2, 3 and 5 (the first three primes)—6, 5, and 6—which code for 0, =, and 0. But a Gödel number must also be an analytic modality, a numeral in the uninterpreted calculus, with the form SSSS…SSS0, if we are to believe that the formalized calculus of arithmetic can describe its own formal properties as a system. (There is no such thing as the prime decomposition of a logical formula, any more than there is the prime decomposition of a circle.) In order for the proof to go through, Gödel numbers must be variously considered as both; the proof cannot renounce its two distinct modes of representation. Their combination via the strategy of Gödel numbering results in a demonstration that exploits and requires a carefully controlled ambiguity. This ambiguity cannot be dispelled; it cannot be dismissed as heuristics but is central to the demonstration; and it goes largely unremarked by Nagel and Newman, whose exposition betrays it nonetheless here and there (see Byers 2010).

In Chap. 1, I noted Karine Chemla's use of the idea of "modalities of application," elicited from her study of the development of projective geometry, focusing on 19th c. projective geometer, Michel Chasles. Chasles pointed to Descartes' method of tangents in analytic geometry as having an enhanced generality not available to the Greeks. Descartes' method of tangents, Chemla notes, is indeed general, but can be applied to curves only when one has succeeded in representing them by an algebraic equation. That is, the 'modality of application' requires that the curve be replaced by an algebraic equation (Chemla 1998). Since this method at first wouldn't work for transcendental curves, precisely because they are not algebraic, many of Leibniz's early efforts in developing the infinitesimal calculus are attempts to apply the method of tangents to non-algebraic curves. The extension of the general methods of analytic geometry to transcendental curves in the long run required novel (non-Cartesian) 'modalities of application' for the same general method.

Logicians are typically very fond of arithmetic. Herbert Enderton, at the beginning of his textbook discussed above, observes that symbolic logic is a mathematical model of deductive thought, and that axiomatic mathematics consists of many logically correct deductions laid end to end: "Thus the deductions made by the working mathematician constitute real-life originals whose features are to be mirrored in our model." He adds that first order logic is "admirably suited to deductions encountered in mathematics. When a working mathematician asserts that a particular sentence follows from the axioms of set theory, he means that this deduction can be translated to one in our model" (Enderton 1972: 2). Logic gets itself into trouble because of its own pretentions: it wants to replace the mathematical languages it formalizes. So, since logical languages must be homogeneous, logic must offer substitutes for (among other items) the natural numbers. We have seen that Enderton's textbook tries to finesse the disparity between logic, set theory, and number theory, as does Nagel and Newman's exposition of Gödel's proof. Ironically, Gödel's proof offers a clear example of the complexity of bringing modes of representation useful for referring, and modes of representation useful for analysis, into rational relation; it is also an example of the fruitfulness of doing so. His solution to the problem is sometimes to let Gödel numbers mean numerals in the logical calculus, when they must play the role of an analytic 'modality of application,' and sometimes to let them mean numbers identified by Indo-Arabic/Cartesian numerals, when they must serve as referents. They must mean both in order for the proof to go through. But then he is exploiting a carefully controlled and fruitful ambiguity, which multiplies the information available to the mathematician, though it is hard for a logician to admit that he is trafficking in heterogeneity.

4 Wiles' Proof of Fermat's Last Theorem Redux

The question, is arithmetic logic, or, can arithmetic be reduced to logic in the sense of Nagel, as I have been arguing, is ill-posed. It confuses areas of research with formal theories. Formal theories are possible elements or aspects of areas of study, and instruments by means of which these areas can be studied. Just as we may learn something by re-constructing a collection of problematic mathematical things as a topological space, so we may learn something by re-constructing a collection of problematic mathematical things as a formal theory. In the latter case, we are treating some of the things of mathematics as if they were discourse; and then we treat discourse as if it were a topic for mathematical investigation, which of course it is: terms, propositions, arguments, sets and categories turn out to have very interesting properties when understood as mathematicial things.

Number theory and logic are two disparate areas of research; for example, they admit extension in very different ways. The solution of problems in number theory is often accomplished by embedding the integers in various kinds of completions, as we saw in the last chapter: the rationals, the reals, the complex numbers, the p-adic numbers, and ultra-products of such completions. The peculiar structure of the integers as discrete units on the one hand makes their construction straightforward, but on the other hand complicates their investigation. Diophantine equations are not assured of solutions because the integers have 'gaps': hence the various strategies for filling those gaps by different completions. Sometimes logical methods can help in this process, but logic has no special claim as a helpmeet, for algebra, complex and real analysis, and topology may also play important roles here. Similarly, the extensions of logic, to second order logic, logics with generalized quantifiers, set theory, recursion theory or model theory, are launched to solve the peculiar problems that logic poses to itself, as it attempts to provide representations for discourse about ever more large and complex mathematical structures. And analysis and topology force developments in logic far beyond those that arithmetic suggests (Grosholz 2007: Chap. 10). The correlations between logic and number theory take a variety of forms, the transfer of information runs both ways, and the role that one area of research plays in the problem-solving strategies of the other is important but episodic and local.

Logicians like Angus Macintyre, Colin McLarty, and Harvey Friedman who analyze Wiles' proof of Fermat's Last Theorem, try to reduce the logical complexity of the proof. In their project of re-writing the proof using discourse of lower logical complexity, certain kinds of abstract structures, used explicitly in Wiles' original proof, may be suppressed, and so aspects of its original organization may be obscured or complicated. Conversely, what was left unremarked or tacit in the original proof, like the 'foundational' justification of some of his sources, are brought to light by the logicians' attempts to re-write the proof; and the methods of approximation they use, as well as their ability to highlight the most combinatoric and arithmetic aspects of the proof, may turn out to be mathematically suggestive to the number theorists. Their articulations can thus be considered as extensions of the original text, where what plays the role of instrument of analysis for the number theorists becomes an object of reference for the logicians. That is, the logicians' reformulations, while they are sometimes intended to replace the original, can more fruitfully be considered as superpositions; the result is then a combination of number theoretical and logical approaches, rationally integrated, so the information available for the solution of various novel problems in both number theory and mathematical logic is increased.

I have been arguing that reasoning in mathematics often generates internally differentiated texts because thinking requires us to carry out two distinct though closely related tasks in tandem, reference and analysis. We investigate things and problems in mathematics because we understand some of the issues they raise but not others; they exist at the boundary of the known and unknown. So too, what plays the role of referent in one mathematical context may appear as an instrument of analysis in another, and vice versa. The philosophical challenge is this: how can a philosopher account for the organization of mathematical discourse, both as it is used in research and as it is used in textbooks? I would like to argue that organization takes place at many levels. It is inscribed in many notations and iconic conventions, in procedures and algorithms, and in the methods that generalize them. It is expressed in the canonization of objects and problems, and the collecting of problems into families. And it is precipitated in iconic as well as symbolic fashion, in diagrams and arrays, as well as in the enunciation of theorems and principles;

moreover, specialized mathematical language must be explained and brought into working relationship by natural language.

Between the principles that govern analysis, and the mathematical things to which we refer, the organization created by the mathematician is quite multifarious. My claim that mathematical objects are problematic (and so in a sense historical and in another sense strongly related to the practices of researchers and teachers) need not lead to skepticism or to dogmatism about the furniture of the mathematical universe; rather, it should lead us to examine the strategies of integration that organize mathematical discourse. We can claim that discourse represents things well without becoming dogmatic, if we leave behind the over-simplified picture of the matching up of reference and analysis as the satisfaction of propositions in a theory by a structure.

Let us look more closely at some aspects of Wiles' proof. Fermat's Last Theorem (1630) states that the equation $x^n + y^n = z^n$, where $xyz \neq 0$, has no integer solutions when n is greater than or equal to 3. Fermat himself proved the theorem for exponent 4, which also reduces the problem to proving the cases where n is an odd prime. Euler produced an (apparently flawed) proof for the case where $n = 3$ (1753), Dirichlet and Legendre simultaneously proved the case where $n = 5$ (1825), and Lamé proved the case where $n = 7$ (1839). Sophie Germaine and Ernst Eduard Kummer produced more general, and generalizable, results in the 19th century, relating the theorem to what would become class field theory in the 20th century (see Bashmakova 1997).

The striking feature of Wiles' proof, to people who are not number theorists, is that it does not seem to be about integers! Here, again, is the opening paragraph of his 108 page paper in the *Annals of Mathematics*: "An **elliptic curve** over **Q** is said to be modular if it has a finite covering by a **modular curve** of the form $X_0(N)$. Any such elliptic curve has the property that its Hasse-Weil **zeta function** has an analytic continuation and satisfies a functional equation of the standard type. If an elliptic curve over **Q** with a given j-invariant is modular then it is easy to see that all elliptic curves with the same j-invariant are modular… A well-known conjecture which grew out of the work of Shimura and Taniyama in the 1950s and 1960s asserts that every elliptic curve over **Q** is modular… In 1985 Frey made the remarkable observation that this conjecture should imply Fermat's Last Theorem. The precise mechanism relating the two was formulated by Serre as the ε-conjecture and this was then proved by Ribet in the summer of 1986. Ribet's result only requires one to prove the conjecture for semistable elliptic curves in order to deduce Fermat's Last Theorem" (Wiles 1995: 443).

This apparent change of referents is explained by the fact that the proof hinges on a problem reduction, just the kind of ampliative problem reduction that interests Carlo Cellucci. The truth of Fermat's Last Theorem is implied by the truth of theTaniyama-Shimura conjecture, that every elliptic curve over **Q** is modular; the converse claim that every modular form corresponds to a certain elliptic curve, as we noted, had already been proved by Eichler and Shimura: Fermat's Last Theorem follows from the two-way correspondence, which rules out counterexamples to it. The condition of modularity is important because then the elliptic curve's ***L*-function** will have an analytic continuation on the whole complex plane, which makes Wiles' proof the first great result of the Langlands Program, and a harbinger of further results. Important problem-reductions combine, juxtapose and even superpose certain kinds of objects (and the procedures, methods and problems typical of them) on other kinds. Wiles' proof is not only about the integers and rational numbers; it is at the same time concerned with much more 'abstract' and indeed somewhat ambiguous and polyvalent objects, elliptic curves and modular forms. So for example at the culmination of Wiles' proof, where analysis has invoked cohomology theory, *L*-theory, **representation theory**, and the machinery of deformation theory, we find the mathematician also involved in quite a bit of down-to-earth number-crunching.

Wiles' proof of Fermat's Last Theorem can be understood in terms of two stages. The first stage was already done for him: it is the result of Eichler-Shimura, which shows that given a certain kind of modular form *f*, we can always find a corresponding elliptic curve E_f (see Eichler 1954; Shimura 1958) The second stage Wiles had to carry out himself, proving the Taniyama-Shimura conjecture, that given a certain kind of elliptic curve *E*, we can always find a certain kind of modular form that renders it 'modular.' Fermat's Last Theorem follows from this correspondence, qualified by a restriction to semi-stable elliptic curves. In the first stage, modular forms are investigated as the objects of reference, and treated 'geometrically' as holomorphic differentials on a certain **Riemann surface**, while elliptic curves are treated as instruments of analysis. In the second stage of Wiles' proof, elliptic curves serve initially as objects of reference (smooth, projective non-singular—without **cusps** or self-intersections—curves of genus 1 over a field *K*, with a distinguished point *O*), while modular forms become the instruments of analysis. One strategy that interests me here is the use of *L*-functions, which figures centrally in the Eichler-Shimura proof, which I discuss in some detail. Another equally important strategy is the use of representation theory in tandem with deformation theory, where ***p*-adic families of Galois representations** figure centrally in the proof of the Taniyama-Shimura conjecture. Had I but world enough and time, I would discuss this too, but then my book would be twice as long (see Frenkel 2013: Chaps. 7–9).

Wiles puts the problem in a very general setting: he finds a universal representation that takes $G_{\mathbf{Q}}$ (the group of **automorphisms** of the algebraic closure of the rationals that leaves the rationals unchanged) to $GL_2(R_\Sigma)$, the set of all 2×2 invertible matrices with entries in the universal deformation ring R_Σ defined with respect to a certain finite set of primes Σ. (Thus, in both cases, there is a hierarchy of representations: there exists a representation to which all the other, more finitary representations can be 'lifted' under the right conditions. Lifting is a kind of generalization specific to deformation theory.) Meanwhile, he also constructs another universal representation: this one takes $G_{\mathbf{Q}}$ to $GL_2(T_\Sigma)$, where T_Σ is the completion of a classical ring of **Hecke operators** acting on a certain space of modular forms. Thus, in both cases, there is a hierarchy of representations: there exists a representation to which all the other, more finitary representations can be 'lifted' under the right conditions; lifting is a kind of generalization specific to deformation theory. Then Wiles shows that T_Σ et R_Σ are isomorphic! This part of the proof, the postulation of universal deformations, might seem, from the point of view of logicians (in particular model theorists concerned with definability), rather extravagant; but in fact this is not where the problem lies for them. Rather, it is Wiles' use of Grothendieck duality about twenty pages earlier (Wiles 1995: 486f.) where the functor categories use universes (see McLarty 2010). I return to this point below. The employment of both strategies is just as much part of the 'analysis' of problems as is the construction of theories.

The problematic nature of reference arises with respect to both elliptic functions and modular forms. In order for the proofs to go through, the mathematicians must specify precisely what kinds of elliptic curves and modular forms they are referring to, in a sufficiently broad manner. But to make the proof practicable, the definition must also be narrowed as much as possible. The delicacy of this issue appears in the first sentences of Wiles' proof. Moreover, a second concern about reference arises here, which is the 'doubling' of the definition of both elliptic curves and modular forms. For example, an elliptic curve is both geometrical (considered as a projective non-singular curve of genus 1, over a field k, with a point O whose coordinates are in k) and algebraic (endowed with a commutative group structure). The specific kind of modular form under investigation also has two aspects. The Eichler-Shimura proof begins with a modular form f from the space of weight 2 cusp forms on $\Gamma_0(N)$ and seeks a corresponding elliptic curve. The latter expression, $\Gamma_0(N)$, is defined as the group of 2×2 integer matrices with determinant 1 that are upper triangular mod N; it is a congruence subgroup of finite index of $SL_2(\mathbf{Z})$, the group of all 2×2 matrices with integer coefficients and determinant 1. (The congruence subgroups of $SL_2(\mathbf{Z})$ are called **modular groups**.) We can identify a **fundamental domain** for the action of the group $\Gamma_0(N)$ restricted to the upper half of the complex plane (the Poincaré complex half-plane) and compactified by adding cusps, which is the curve $X_0(N)$: it is a **Riemann surface** (We define f to be holomorphic on this Riemann surface; a cusp form, which is a kind of modular form, vanishes at the cusps).

Because the expression $f(z)d(z) = f\left(\begin{pmatrix} a & b \\ c & d \end{pmatrix} z\right) d\left(\begin{pmatrix} a & b \\ c & d \end{pmatrix} z\right)$ is invariant under the action of all the $\begin{pmatrix} a & b \\ c & d \end{pmatrix}$ in $\Gamma_0(N)$ on the complex plane, this gives $f\left(\frac{az+b}{cz+d}\right) = (cz+d)^2 f(z)$ and in particular $f(z+b) = f(z)$ for all integers b. Then f has a Fourier expansion in powers of $q = e^{2\pi i z}$, namely $f(z) = \sum_{n=1}^{\infty} a_n q^n$. Here is the re-doubling of the definition of a modular form: we identify the space of weight 2 cusp forms on $\Gamma_0(N)$, the $f(z)$, with the space of holomorphic differentials on the Riemann surface $X_0(N)$, the $f(z)dz$. We can define Hecke operators T_m on the space of these modular forms, which satisfy certain important properties and form a commutative algebra T under composition; we also define a Petersson inner product $\langle , \rangle$, so that all f and g in this space satisfy the relation $\langle T_m f, g\rangle = \langle f, T_m g\rangle$. We can then study this space in terms of the eigenfunctions of the T_m, and this study allows us to define an important class of modular forms called 'newforms.' The newforms and their suitable transforms constitute a basis for the weight 2 cusp forms for $\Gamma_0(N)$, and the elements of this basis are common eigenfunctions of all Hecke operators. For each newform *f*, in its Fourier expansion

$$f(z) = \sum_{n=1}^{\infty} a_n q^n$$

we can assume $a_1 = 1$, so that all the coefficients a_n are eigenvalues of the Hecke operators T_n and each a_n lies in **Z**. The study of modular forms is thus reduced to the study of newforms.

L-functions are, again, generalizations of the Riemann zeta function, and the Dirichlet series, representations of functions in terms of infinite series that give valuable information about the function even at values where its infinite series does not converge. In the case of these modular forms, we have two ways of expressing their *L*-functions, one as an infinite sum, and the other as an infinite product.

$$L(s,f) = \sum_{n=1}^{\infty} a_n n^{-s} = \prod_{p \text{ does not divide } N} \frac{1}{1 - a_p p^{-s} + p^{1-2s}} \prod_{p \text{ divides } N} \frac{1}{1 - a_p p^{-s}}.$$

The Eulerian of the *L*-function at all primes *p* reflects the fact that the Hecke algebra T is generated by the T_p (*p* prime). To an elliptic curve there is also an associated *L*-function with a form similar to that given above, arising from counting points. So for a given *f* with integral coefficients a_n we want to find a suitable elliptic curve *E* whose *L*-function coefficients b_n will match the *L*-function coefficients a_n of *f*. But first, we must return to the definition of an elliptic curve, in the hope that, just as the *L*-functions attached to modular forms have nice properties, the *L*-functions attached to elliptic curves will have equally nice, informative properties.

Modular forms belong to complex analysis, though they were originally intended to help solve problems in number theory. Elliptic curves belong to algebraic geometry. As noted above, they are defined in a double sense. Taken geometrically, an elliptic curve is a smooth, projective non-singular (with no cusps or self-intersections) curve of genus 1, over a field *k*, with a point *O* with coordinates in *k*. Taken algebraically, it is a projective variety endowed with a group structure, that is, an operation which is necessarily commutative with the point *O* as the identity element. The Riemann-Roch theorem tells us that every elliptic curve defined over a field with characteristic not 2 or 3 is given by some generalized Weierstrass equation,

$$(^*) y^2 = x^3 + ax + b.$$

Elliptic curves reveal different features, depending on what field they are taken over. Over the complex numbers, an elliptic curve is described as the complex plane modded out by a lattice, making it a torus. This is a Riemann surface with a group structure inherited from the addition of complex numbers. To see it geometrically, note that the meromorphic functions on E, which is a torus, are doubly periodic. It turns out that these elliptic functions form a field generated by two functions x and y which, with some complex numbers a and b, satisfy the relation (*) given above. Over the real numbers it is a plane curve, and if we add a point at infinity as the distinguished point O, it becomes the projective version of this curve. The points on E with rational coordinates form a discrete, finitely generated Abelian group. The L-function of an elliptic curve with **conductor** N is defined in the following way.

$$L(E,s) = \prod_{p\,does\,not\,divide\,N} \frac{1}{1 - b_p p^{-s} + p^{1-2s}} \prod_{p\,divides\,N} \frac{1}{1 - b_p p^{-s}}.$$

The Eichler-Shimura result shows that when we carry out the construction mentioned above, namely, given a weight 2 cusp form f for $\Gamma_0(N)$ with integral coefficients a_n, the **Jacobian** $J_0(N)$ of $X_0(N)$ contains a special subvariety A so that $J_0(N)/A$ will be the elliptic curve sought, whose L-function has coefficients b_n that match the L-function coefficients a_n of the original modular form f (For a more detailed account of this proof, which I have only summarized here, please see Appendix B).

In sum, the Jacobian of a modular curve is analogous to a complex elliptic curve in that both are complex tori and so have Abelian group structure. Every complex elliptic curve with a rational j-invariant is the holomorphic homomorphic image of a Jacobian. Indeed, the elliptic curve is the image of a quotient of a Jacobian, the Abelian variety associated to a weight 2 **eigenform** (a cusp form) f. Only weight 2 eigenforms with rational Hecke eigenvalues correspond to elliptic curves; more general eigenforms correspond to Abelian varieties. All the ways of explaining this correspondence involve appeals to 'large' structures to which the rather modest newform $f(z)$ can be lifted, investigated, and then re-deposited with a new affiliation, to an elliptic curve. And this is just a sketch of the first part of the proof; the more difficult part was Wiles' proof of the Taniyama-Shimura conjecture, showing that given a certain kind of elliptic curve E, we can always find an associated modular form f.

As I have argued above, mathematical discourse must carry out two distinct tasks in tandem, analysis and reference. In the case of number theory, the referents are integers and rational numbers in one sense and additionally, in a broader sense given the problem reduction at the heart of Wiles' proof, modular forms and elliptic curves. For logic, the referents are propositions and sets (and perhaps also formal proofs), or, if we include the broader range of category theory as part of logic, categories (and perhaps also functors). Thus what is an aspect of analysis for the number theorist is an aspect of reference for the logician. Moreover, techniques of calculation that preoccupy the number theorist remain tacit for the logician because they directly involve numbers, and considerations of logical complexity that concern the logician remain tacit for the number theorist because they are not conditions of solvability for problems about numbers. This disparity is inescapable, but it is also positive for the advance of mathematics. For when what remains tacit in one domain must be made explicit in another in order to bring the domains into rational relation, novel strategies of integration must be devised.

5 McLarty and Friedman

A notable feature of Andrew Wiles' proof of Fermat's Last Theorem is that it invokes cohomology theory (inter alia) and thus Grothendieck's notion of successive universes, which from the point of view of set theory become very large; and yet the detail of the proof stays on relatively low levels of that vast hierarchy. In a recent essay, Colin McLarty offers foundations for the cohomology employed in Wiles' proof at the level of finite order arithmetic; he uses Mac Lane set theory, which has the proof theoretic strength of finite order arithmetic, and Mac Lane type theory, a conservative extension of the latter (McLarty 2010). Angus Macintyre is re-working aspects of the proof (bounding specific uses of induction and comprehension) to bring it within a conservative *n*th order extension of Peano Arithmetic (Macintyre 2011) and Harvey Friedman has informally speculated that it could be further reduced to Exponential Function Arithmetic.

Meanwhile, the significant re-working and extension of the proof by number theorists proceeds independently of logic, in the sense that number theorists don't seem particularly concerned about the logical complexity of their methods. (See for example recent work by Christophe Breuil, Brian Conrad, Frederick Diamond, Mark Kisin, and Richard Taylor.) On the one hand, we see number theorists choosing logically extravagent methods that usefully organize their investigations into relations among numbers, as well as elliptic curves, modular forms, and their *L*-functions, inter alia, and make crucial computations visible and possible. On the other hand, we see logicians analyzing the discourse of the number theorists, with the aim of reducing its logical complexity. Should number theorists care whether their abstract structures entail the existence of a series of strongly inaccessible cardinals? Serre and Deligne, for example, do sometimes seem to be concerned about the logical complexity of their methods (Macintyre 2011: 10). Will the activity of logicians produce useful results for number theorists, or is it enough if they answer questions of interest to other logicians, such as whether in fact Fermat's Last Theorem lies beyond the expressive strength of Peano Arithmetic (and thus might be a historical and not merely artificially constructed example of a Gödel sentence)?

As I have argued above, mathematical discourse must carry out two distinct tasks in tandem, analysis and reference. I use as an illustration of this disparity and the possibility of productive integration the work of Angus Macintyre (a model theorist) and Colin McLarty (a category theorist) on Wiles' proof of Fermat's Last Theorem. At issue is Wiles' use of Grothendieck cohomology, as set forth in various writings and editions of *Eléments de Géométrie Algébrique* over the third quarter of the 20th century. Colin McLarty writes that Grothendieck pre-empted many set theoretic issues in cohomology by positing a universe: a set large enough that the habitual operations of set theory do not go outside it. His universes prove that ZFC is consistent, so ZFC cannot prove that they exist. (McLarty 2010: 359–361). Wiles invokes Grothendieck cohomology and by implication the vast universes it involves around page 486 of Wiles' "Modular elliptic curves and Fermat's Last Theorem," where he uses Grothendieck duality and parts of Mazur's work, and the textbook *An Introduction to Grothendieck Duality* by Altman and Kleiman (see Mazur 1977; Altman and Kleiman 1970). The path through these books leads back to Grothendieck's *Eléments de Géométrie Algébrique* and functor categories that use universes (Grothendieck and Dieudonné 1960–1967).

As McLarty points out, the odd thing is that these rather oblique and vague references are all that Wiles offers the logician-reader interested in tracing back his assumptions to their origins; indeed, Wiles never offers an explicit definition of cohomology. McLarty speculates that Wiles may be assuming that the Anglophone reader will consult the standard textbooks, Hartshorne's *Algebraic Geometry* or Freyd's *Abelian Categories*; but these books are not included in the extensive references at the end of Wiles' article (Hartshorne 1977; Freyd 1964). In any case, both Hartshorne and Freyd treat questions of proof and foundations in a rather cavalier manner. McLarty writes that Hartshorne quantifies over functors between categories which are not well defined in ZF, and he does not prove the basic results he uses. He cites Freyd's *Abelian Categories* for proofs and also sketches several other strategies one could use. Freyd in turn waves off the question of foundations by claiming he could use some theory like Morse-Kelley set theory, a non-conservative extension of ZF. And that is true of his chief results (though at least one of his exercises goes beyond that). In general, McLarty argues, from the point of view of the logician, Wiles proves no theorems from the ground up (McLarty 2010: 367–368).

Wiles makes use of cohomology theory, and the deformation theory allied with it, because it helps him to organize the results he needs for his proof; but it is not what his proof is *about*. The logical strength of the theory does not really concern him, so he lets it remain for the most part tacit and unanalyzed in his exposition. For logicians concerned with model theory, or with the meaning of Gödel's Incompleteness Theorems, however, the logical strength of Wiles' proof of the Taniyama-Shimura conjecture, or of other proofs still to be discovered that are now emerging from it, is paramount. It must be made explicit, in order to explore the possibility of proofs of the same result but with lower logical complexity. One way of posing this question, however, leads us back to the discussion of the double nature of the definitions in Wiles' proof in the preceding sections. Can a proof in number theory really do without geometry? This is a central question because, even if one succeeds in wresting large parts of cohomology theory into first or second order arithmetic, even second order arithmetic will not provide any uncountable fields like the reals or complex numbers or p-adic numbers. An appropriate formalization will apply to them if one assumes they exist, but will not prove they exist. So we are dealing not only with the disparity between number theory and logic, but also with the disparity between number theory and geometry.

6 Macintyre

In his essay, "The Impact of Gödel's Incompleteness Theorems on Mathematics," Angus Macintyre begins by noting the positive contributions of logicians to research in various branches of mathematics (apart from mathematical logic itself). He cites Presburger's work on the ordered Abelian group $\mathbb{Z}$, which underlies much of p-adic model theory; Tarski's work on real closed fields; the uses of Ramsey's Theorem in harmonic analysis; and Herbrand and Skolem's contributions to number theory: Herbrand to ramification theory and cohomology, and Skolem to p-adic analytic proofs of Finiteness Theorems for Diophantine equations (Macintyre 2011: 3–4). He then summarizes the reactions of number theorists to Gödel's Incompleteness Theorems. "In the last thirty-five years, number theory has made sensational progress, and the Gödel phenomenon has surely seemed irrelevant," even though number theorists are sensitive to the effectivity or logical complexity of their results. On the one hand, artificially constructed statements that are formally undecidable seem to be mathematically uninteresting: "the equations whose unsolvability is equivalent (after Gödel decoding) to consistency statements have no visible structure, and thus no special interest." On the other hand, the really important results seem mostly to be decidable, at least in principle: "there is not the slightest shred of evidence of some deep-rooted ineffectivity." (Macintyre 2011: 4–5).

Macintyre observes further that while logic is sometimes a good idiom for recasting mathematical research (as in the cases given above), sometimes it uncovers results that are of interest to logicians, but not to geometers or number theorists. What model theory reveals, generally speaking, are the natural "logical" or arithmetic-combinatorial features of a subject matter or problem context. Even when the subject matter is squarely geometrical or topological, these features may be important; but we cannot expect them to tell the whole story. Logic seems more apt for the work of analysis than the work of reference in other mathematical domains. For example, discussing the work of C.L. Siegel, Macintyre writes, "A propos the decision procedure for curves, the natural logical parameters of the problem, such as number of variables, and degree of polynomials involved, obscure the geometrical notions that have proved indispensable to much research since … 1929… If one's formalism obscures key ideas of the subject, one can hardly expect logic alone to contribute much." (Macintyre 2011: 6).

In the Appendix to the paper, Macintyre sketches his conception of what a re-writing of Wiles' proof might look like, if it were carried out by a logician who wanted to show that there is no need for strong second-order axioms with existential quantifiers involved; in other words, he asks what the proof might look like if confined within first-order Peano Arithmetic (PA). Macintyre's conjectured re-writing breaks the proof up into a series of 'local issues,' giving arithmetic interpretations of specific parts of real, complex or p-adic analysis or topology. He points out that zeta functions, L-series and modular forms are all directly related to arithmetic: "There is little difficulty in developing the basics of complex analysis for these functions, on an arithmetic basis, sufficient for classical arithmetical applications… nothing would be gained by working in a second-order formalism, in a very weak system. At best such systems codify elementary arguments of general applicability." (Macintyre 2011: 7). Thus for the number theorist interested in advancing research by generalization, the re-writing of the logician would not be of immediate interest; but for the logician, the re-writing is of central importance.

However, since number theorists are in fact concerned about the logical complexity of their methods, in retrospect they would be motivated at least to study a first-order version of the proof, even if from the perspective of their immediate interests it appears over-detailed and oddly arranged. McLarty observes, "Macintyre points out that analytic or topological structures such as the p-adic, real and complex numbers enter Wiles's proof precisely as completions of structures such as the ring of integers, or the field of rational numbers, which are interpretable in PA [Peano Arithmetic]. Macintyre outlines how to replace many uses of completions in the proof by finite approximations within PA. He shows how substantial known results in arithmetic and model theory yield approximations suited to some cases. He specifies other cases that will need numerical bounds which are not yet known. Theorems of this kind can be very hard. He notes that even routine cases can be so extensive that 'it would be useful to have some metatheorems.'" (McLarty 2010: 363).

From the point of view of number theory, this re-writing would damage the organization and perspicuity of the original proof. Thus, the 'logically extravagent' general methods and theorems seem to be needed to organize the practice of number theorists. However, the 'messy and piecemeal' methods of the logician reveal aspects of the proof (its reducible logical complexity) central to the research practice of logicians. An analyst or topologist need not be interested in replacing **R** or **C** by 'finite approximations within Peano Arithmetic,' but a model theorist is highly motivated to spend quite a bit of time and ink in the attempt. Mathematicians like Wiles, Ribet and Mazur posit the big structures to set their problem in the best conceptual framework possible, so they can see how to solve the problem and then how to generalize the result; model theorists like Macintyre break the big structures into smaller approximations, in order to solve different kinds of problems. Neither one thinks that a finite approximation is identical to the original object it approximates; but for different reasons, and for specified purposes, locally, the number theorist and the model theorist are both willing to entertain the conjectural equivalence. The reduction is both too weak and too strong, but it is useful.

Macintyre is concerned about the way many people misunderstand the import of Gödel's incompleteness results, and overstate the inability of logic to capture the content of important theorems. (It is useful to weigh his arguments against those of Carlo Cellucci, discussed in Chap. 3.) So at least part of what he is trying to do in the Appendix is to show that the 'logical,' that is, the arithmetic-combinatorial, aspects of e.g. Wiles' proof loom very large, and can be captured and re-stated perspicuously by logicians (in particular, model theorists). I would observe that the canonical objects of geometry and topology can typically be treated by arithmetical-combinatorial approaches, even if those approaches do not allow us to capture the canonical objects categorically, or to prove their existence. The work of logic in other mathematical domains is not reference but analysis. Macintyre also points out that the 'monsters,' the sentences, functions, or set-theoretical objects that seem to be squarely beyond the realm of the effective, seem (so far) not very interesting to mathematicians working in other areas. One can point to them, but there doesn't seem to be much to say about them. Like inaccessible ordinals, their very inaccessibility makes them mathematically inert and unrelated to the items and methods that currently drive research. Thus logicians may have a great deal to teach number theorists (and geometers, and set theorists) about the tacit assumptions that guide their choices about what things, procedures, and methods to canonize; and the interaction between logic and number theory, for example, may give rise to novel objects, procedures and methods still to be discovered.

References

Altman, A., & Kleiman, S. (1970). *An introduction to Grothendiek duality theory*. New York: Springer.

Bashmakova, I. (1997). *Diophantus and Diophantine equations*. Washington D. C: Mathematical Association of America.

Byers, W. (2010). *How mathematicians think: Using ambiguity, contradiction and paradox to create mathematics*. Princeton: Princeton University Press.

Carnap, R. (1928). *Der logische aufbau der welt*. Hamburg: Meiner.

Carnap, R. (1983). *The logical structure of the world*. R. A. George (Trans.). Berkeley: University of California Press.

Chemla, K. (1998). Lazare Carnot et la généralité en géométrie. *Revue d'histoire des mathématiques, IV*, 163–190.

Churchill, R. P. (1990). *Logic: An introduction*. New York: St. Martin's Press.

Eichler, M. (1954). Quaternäre quadratische Formen und die Riemannsche Vermutung für die Kongruenzzetafunktion. *In Archiv für mathematische Logik und Grundlagenforschung, 5*, 355–366.

Enderton, H. B. (1972). *A mathematical introduction to logic*. New York: Academic Press.

Frenkel, E. (2013). *Love and math: The heart of hidden reality*. New York: Basic Books.

Freyd, P. (1964). *Abelian categories*. New York: Springer.

Gödel, K. (1931). Über formel unentscheidbare Sätze der Principia Mathematica und verwandter Systeme. *Monatshefte für Mathematik und Physik, 38*, 173–198.

Gödel, K. (1992). *On formally undecidable propositions of* Principia Mathematica *and related systems*. B. Meltzer (Trans.) New York: Dover.

Grosholz, E. (2007). *Representation and productive ambiguity in mathematics and the sciences*. Oxford: Oxford University Press.

Grothendieck, A., & Dieudonné, J. (1960–1967). *Eléments de géométrie algébrique*. Paris: Institut des Hautes Etudes Scientifiques.

Hartshorne, R. (1977). *Algebraic closed fields*. New York: Springer.

Hendry, R. (2001). Mathematics, representation, and molecular structure. In U. Klein (Ed.), *Tools and modes of representation in the laboratory sciences* (pp. 221–236). Dordrecht: Kluwer.
Macintyre, A. (2011). The impact of Gödel's incompleteness theorems on mathematics. *Kurt Gödel and the foundations of mathematics: Horizons of truth* (pp. 3–25). Cambridge: Cambridge University Press.
Mazur, B. (1977). Modular curves and the Eisenstein ideal. *In Publications Mathématiques, Institut des Hautes Études Scientifiques, 47*, 133–186.
McLarty, C. (2010). What does it take to prove Fermat's last theorem? Grothendieck and the logic of number theory. *Bulletin of Symbolic Logic, 16*(3), 359–377.
Nagel, E., & Newman, J. (1958). *Gödel's proof.* New York: New York University Press.
Shimura, G. (1958). Correspondances modulaires et les fonctions ζ de courbes algébriques. In *Journal of the Mathematical Society of Japan, 10*, 1–28.
Van Fraassen, B. (1989). *Laws and symmetry*. Oxford: Oxford University Press.
Van Fraassen, B. (2008). *Scientific representation: Paradoxes of perspective*. Oxford: Oxford University Press.
Wagner, R. (2008). Post-structural readings of a logical-mathematical text. *Perspectives on Science, 16*(2), 196–230.
Wiles, A. (1995). Modular elliptic curves and Fermat's last theorem. In *Annals of Mathematics, 142*, 443–551.

Chapter 6
The Representation of Time in the 17th Century

Augustine of Hippo once wrote, "What then is time? There can be no quick and easy answer, for it is no simple matter even to understand what it is, let alone find words to explain it. Yet, in ordinary conversation, no word is more familiarly used or more easily recognized than 'time.' We certainly understand what is meant by the word both when we use it ourselves, and when we hear it used by others. What, then, is time? I know well enough what it is, provided that nobody asks me; but if I am asked what it is and try to explain, I am baffled" (Augustine 1961: XI.14.17, 263–264). Here Augustine, without precedent, makes time itself a topic of philosophical reflection, as Descartes makes consciousness itself (apart from the objects of consciousness) a topic of philosophical reflection in *Meditations* I and II. Augustine realized that discourse, and indeed in a sense lived experience, retrieve things from the asymmetrical flow of temporality. We experience for the most part a world of stable objects, and our language assigns permanent names to them, their salient properties and relations, and their characteristic functions and activities. The names of Gods are concepts, as the classicist James Redfield has often noted; so too concepts are little gods, conferring immortality on the things they organize. That immortality is in part illusory, because things persist stably only for awhile, and the gods seem to have departed; but it is also the key to scientific knowledge because that stability is real. We live in an organized world, not a maelstrom.

We now have the habit, learned in the 20th century from Relativity Theory, to speak of 4-dimensional space-time; yet I would argue (with Bergson) that time is not like space, and we should be wary of the assimilation. In Classical Mechanics, space makes physical symmetry possible; it allows for symmetries and indeed the study of symmetries in geometry, physics and chemistry is a way of obtaining information about the structure of space. Time, by contrast, is a principle of asymmetry, if one is willing to assert the reality of its directedness. Bas van Fraassen writes that periodicity is a certain kind of symmetry in time (Van Fraassen 1989/2003: 252). While this is a suggestive formulation, it is wrong in one obvious sense. If I decide to leave my office right now, and go out for an afternoon promenade, I can walk north or south, east or west, and if I can find some stairs or

E.R. Grosholz, *Starry Reckoning: Reference and Analysis in Mathematics and Cosmology*, Studies in Applied Philosophy, Epistemology and Rational Ethics 30, DOI 10.1007/978-3-319-46690-3_6

an elevator I can go up and down. But I cannot decide to visit this morning. On my circuit, I can walk up the library steps and then down them later on the way back to my office; but I cannot return to this morning. Space allows for the possibility of symmetry in our displacements, to and fro; time appears to allow for symmetry (in the guise of periodicity) but only because of its alliance with space, and the systems that organize themselves in space. Tomorrow morning, it will be the same clock-time as it is now, and the hands on my clock will point to the same numbers as they do now; but it will be tomorrow. Our ways of keeping time, dependent on the periodicities of the solar system, allow the clock face to be round and the calendar pages to be square, mapping asymmetrical, unrepeating time back on itself by means of finite spatial arrays.

What then is time? Important scientific texts dealing authoritatively with time often occur in the midst of debates that a philosopher might want to call dialectical. Thus Newton argues with Leibniz; the founders of thermodynamics are countered by Boltzmann; the defenders of General Relativity Theory oppose some (though not all) of the proponents of Quantum Mechanics; and contemporary cosmologists find themselves at odds. The structure of these debates in every case strikes me as similar. One side takes on the abstract, more discursive project of theorizing, what Leibniz called analysis; here time is treated as a condition of the intelligibility of things and expressed mathematically. The other side adopts more concrete, often materially realized strategies for achieving successful reference; here time is treated dynamically as a feature of physical systems, including clocks. My suggestion is that if we pay attention to how these opposing discourses are subsequently integrated, and why the polarity of analysis and reference nonetheless always reasserts itself, we will gain some insight into the nature of time.

1 Galileo's Representation of Time

Theorem I, Proposition I in the section 'Naturally Accelerated Motion' in the Third Day of Galileo's *Discorsi* (*Two New Sciences*) states: "The time in which any space is traversed by a body starting from rest and uniformly accelerated is equal to the time in which that same space would be traversed by the same body moving at a uniform speed whose value is the mean of the highest speed and the speed just before acceleration began" (Galilei 1914/1954: 173–174). The accompanying figure has two components, a vertical line CD on the right representing space traversed (not just distance but displacement), and a two-dimensional figure AGIEFB on the left, in which AB represents time (Fig. 1). The two-dimensional figure reproduces Oresme's diagram that applies the important theorem reached by the logicians at Merton College, Oxford, concerning the mean value of a 'uniformly difform form' to uniformly accelerated motion. However, Galileo rotates the diagram by 90° because he is going to apply it even more specifically to the case of free fall, and wants to emphasize its pertinence to the vertical trajectory CD. Koyré points out that the genius of this set of figures is that AB represents not the distance traversed,

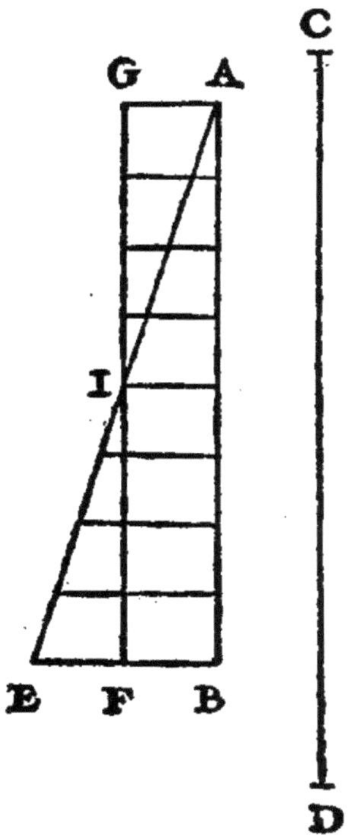

Fig. 1 Galileo, *Discorsi*, Third Day, Naturally Accelerated Motion, Theorem I, Proposition I

but *time*, for Galileo like Oresme has wrested geometry from the geometer's preoccupation with extension and put it in the service of mechanics, whose objects of study are dynamic and temporal (Koyré 1939: 11–46). The parallels of the triangle AEB perpendicular to AB stand for velocities, and the area of the triangle as a whole, taken to be a summation of instantaneous velocities, therefore represents distance traversed. Thus the triangle AEB exhibits the way that uniformly increasing velocity and time are related in the determination of a distance. Visual inspection of this diagram shows that, with each passage of n units of time, the area of triangle AEB increases in the following way: for n number of terms,

$$1 + 3 + 5 + 7 + \ldots [2n - 1] = n^2$$

that is, distance traversed is proportional to the square of the time elapsed.

Galileo spells this out in a more indirect and rigorous way in his Theorem II, Proposition II, which characterizes free fall. Here the vertical line acquires a numerical and literal articulation: the points H, L, M, N and I mark the passage of successive equal intervals of time, and the units of length inscribed on the line indicate that in the first interval of time, the body falls 1 unit, in the second 3, in the third 5, in the fourth 7, and so on. Thus we see depicted by this line, as by the earlier triangle, that distance traversed is proportional to the square of the time elapsed

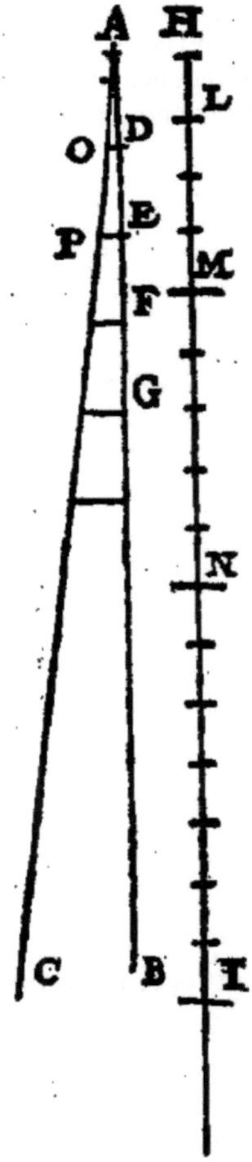

Fig. 2 Galileo, *Discorsi*, Third Day, Naturally Accelerated Motion, Theorem II, Proposition II

(Fig. 2) (Galilei 1914/1954: 174–175). Then, in the Fourth Day, Theorem I, Proposition I, he combines his earlier analysis of inertial motion with his analysis of free fall, and produces his celebrated diagram that proves that the trajectory of a projectile is parabolic, the combination of uniform motion along (what we would call) an x-axis, and of uniformly accelerated motion along a y-axis (Fig. 3). The line *abcde* represents time as well as the displacement of a body in inertial motion (since, as Galileo proved earlier, in such motion the intervals of time elapsed are proportional to the intervals of distance traversed); the intervals *bc*, *cd*, *de* are equal. The line *bogln*, borrowed from his earlier analysis of free fall, is ruled off in intervals proportional to 1, 3, 5, 7, and so forth. The genius of the diagram is the perpendicular juxtaposition of line *bogln* with *abcde*, which displays Galileo's insight that projectile motion is "compounded of two other motions, namely, one uniform and one naturally accelerated," so that its path is a semi-parabola (Galilei 1914/1954: 248–250) (see also Grosholz 2007: Chap. 1, 2011a).

Thus in this diagram, time is represented in two different ways. In line *abcde*, time is 'analytic' or geometrical, and *t* is proportional to the distance traversed. The units of length can just as well be taken to be units of time, so that time can be specified independently of the dynamical event. In line *bogln*, however, time is 'referential,' in the sense that it is a function of the dynamical event. Distance traversed is proportional to the square of time, so that *t* is proportional to the square root of the total distance traversed: the reckoning of time is tied to the dynamical event of the falling body. The mathematics that produced this line (as we saw in Figs. 1 and 2) is a foray into methods that a later generation will call the differential and integral calculus. We read the passage of time from the body as it falls, and that reckoning points towards the discovery of the gravitational constant, an empirical feature of free fall, as well as, later on, the formulation of gravity as a force. Compose the 'analytic' and 'referential' records of time in the diagram, and we see the parabolic path of the projectile.

2 Newton's Representation of Time

Newton's proof of the inverse square law in Proposition XI of Book I of the *Principia* also depends on the combination of three different ways of representing time. We can locate them by tracing Newton's argument through Book I: The Motion of Bodies. The reasoning begins with Proposition I of Book I, Newton's generalization of Kepler's law of areas: "The areas which revolving bodies describe

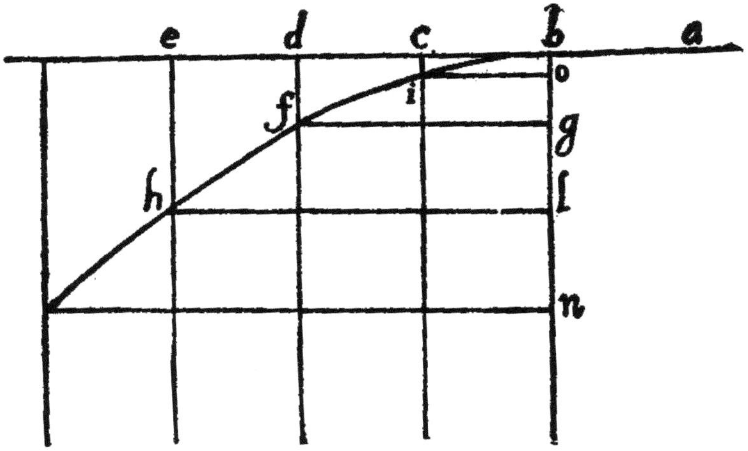

Fig. 3 Galileo, *Discorsi*, Fourth Day, The Motion of Projectiles, Theorem I, Proposition I

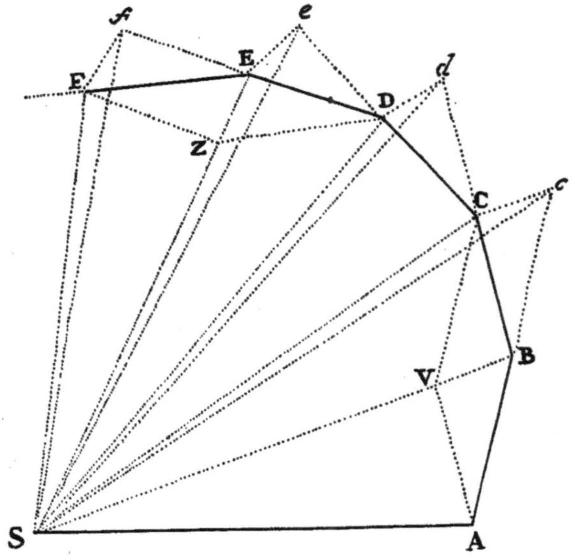

Fig. 4 Newton, *Principia*, Book I, Section II, Proposition I, Theorem I

by radii drawn to an immoveable center of force do lie in the same immoveable planes, and are proportional to the times in which they are described" (Fig. 4) (Newton 1934: I, 40–42; see also De Gandt 1995). In the diagram, S is the center of force; the body proceeds on an inertial path from A to B, and would continue straight on to c if it were not deflected by an 'impulse' of force so that it arrives at C. Then cC (= BV) represents the deflection of the body due to force; the perimeter ABCDEF… is the trajectory of the body as it is deflected at the beginning of each equal interval of time by discrete and instantaneous impulsions from S. Given Euclidean results about the equality of triangles, we can read off the diagram that equal areas are described in equal times. Thus, time is directly proportional to the line AB (because the motion of the body is inertial) and to the area ΔSBC; and the square of time is directly proportional to cC in the first moment of free fall.

In Proposition VI, where Newton gives a general definition of the action of a central force, time is directly proportional to PR, the inertial path that the revolving body would follow if it were not deflected by the central force. This is the 'analytic', geometrical representation of time that we recall from Galileo's diagram of projectile motion. Time is also directly proportional to the area ΔSPQ [which in the limit is equal to ½ (SP × QT)], by Newton's rewriting of Kepler's Law; and QR is directly proportional both to the force F and to the square of time, by Newton's adaptation of Galileo's model of free fall "in the first moment of fall." Here time is defined 'referentially,' in terms of the dynamical nexus of the body and the center of force, and the magnitudes are 'evanescent,' to use the vocabulary of Newton's version of the infinitesimal calculus. Thus QR is proportional to the product of the force F and the square of (SP × QT), so that (rearranging the proportion) F is inversely proportional to ($SP^2 \times QT^2/QR$) (Fig. 5) (Newton 1934: I, 48–49).

Fig. 5 Newton, *Principia*, Book I, Section II, Proposition VI, Theorem V

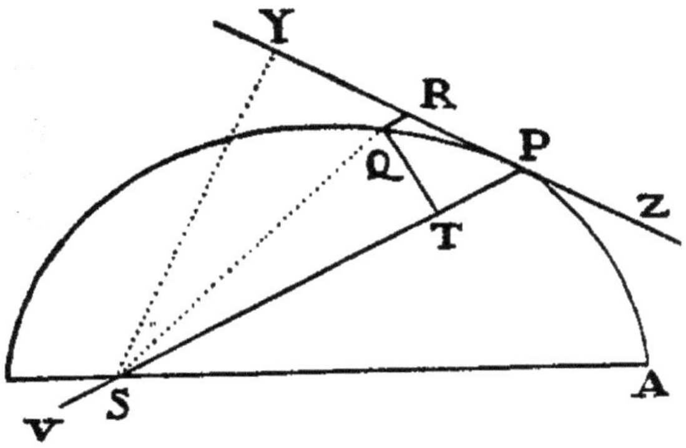

In the subsequent propositions leading up to Proposition XI, the proof of the inverse square law, Newton investigates other situations suggested by the generality of Proposition VI. He finds the law governing the center of force when the body moves in a circle and the center of force is at the center of the circle; when the body moves in a circle and the center of force is very far removed; when the body moves in a spiral and the center of force is in the center of the spiral; and when the body moves in an ellipse and the center of force is in the center of the ellipse, not at one of the foci.

Finally, Newton finds the law governing the center of force when the body moves in an ellipse and the center of force is located at one of the two foci. To apply his general result to the particular case of an elliptical trajectory, Newton makes use of the specific geometrical attributes of ellipses, and shows that in this case the central force F must be inversely proportional to the square of the distance, because the formula ($SP^2 \times QT^2/QR$) reduces to SP^2, the square of the distance from the body to the center of force (Fig. 6). I will not rehearse the details of the proof here as I do in Chap. 7 of my *Representation and Productive Ambiguity in Mathematics and the Sciences* (see Grosholz 2007: Chap. 7, 2011a). However, the main point is that the proof goes through because the three distinct ways to define time (with respect to Descartes' definition of inertial motion, Kepler's law of areas, and Galileo's analysis of free fall) are usefully combined in the proportions, which carefully segregate the geometrical elements from the evanescent and dynamical elements of the diagram, only to recombine them at the end in order to arrive at the celebrated result.

However, a tension remains inside this work of combination, because the modes of representation are disparate; one cannot be reduced to the other, and in certain situations they are clearly not equivalent. Here is a simple example. If we change the inertial frame in which we set the projectile motion analyzed by Galileo so that the inertial frame moves at the same rate as the horizontal component of the projectile's motion, time will not be represented by a line but by a point (since the projectile's horizontal displacement will be zero) and the parabola will become the line *bogln*, the line of free fall. Then time as geometrically represented will

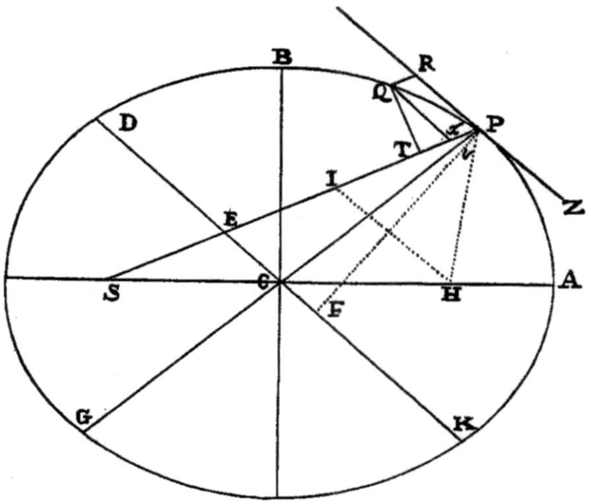

Fig. 6 Newton, *Principia*, Book I, Section II, Proposition VI, Theorem V

be proportional to zero, given the convention by which Galileo has chosen to represent it. Time as dynamically represented, however, will be finite and growing; so they are not equivalent. In fact, Newton's absolutist doctrine of space and time, as we have already seen, is too strong; it makes all motion absolute, so that the physical equivalence of rest and inertial motion, and of motion in different inertial frames, is lost. But the equivalence, on which depends the 'Galilean invariance' of physical laws, is central to his mechanics. This disparity points to tensions within mechanics that eventually lead to the special and general theories of relativity. Moreover, geometry and infinitesimal analysis remain distinct mathematical domains. Despite the obvious and extensive overlap, geometry retains its own items, modes of representation, and methods, quite different from those of infinitesimal analysis, and gives rise to different kinds of offspring, like projective geometry. And while the ideal items and processes of classical mechanics remain central to the development of analysis as the home of transcendental as well as algebraic curves and of the differential equations, they do not exhaust it, as it veers off on its own towards the infinitary.

3 Leibniz on Method

Leibniz's conception of method, dependent on his metaphysics, tends to make the study of history scientific and the scientific study of nature historical. This notable effect of his method indicates that both his method and metaphysics are intimately related to time; that is, his treatment of temporality should provide an important key to his method and metaphysics, and thus to his work as scientist and historian. In both domains, a Leibnizian savant must engage in analysis, the search for the conditions of intelligibility, the requisites, of what exists; this search is ampliative, leading from the simple schema 'S is P' to series and networks of relations. The doctrine of pre-established harmony, and conversely the doctrine of monadic expression, underwrites the movement from 'S is P' to relational idioms, like algebraic formulas, infinite series, differential equations, and combinatorial schemes, or deductive inference forms, probabilities and trees, which apply to a variety of different kinds of objects and operations. As I have been arguing, the method of analysis, uncovering a condition of intelligibility P of an existent, S, is ampliative; it discovers what is primitive, or fundamental, or 'simple' in the investigation of a complex thing, and this investigation, by making the implicit explicit, uncovers the general or canonical form of the thing, the general formula according to which it can be treated systematically. Things that exist are not just non-contradictory, but completely—indeed infinitely—conditioned, governed both by the principle of contradiction and the Principle of Sufficient Reason; so the systematicity uncovered by analysis is not illusory or superficial but rather well-founded. To understand something, a Leibnizian savant must not only witness its internal consistency, but also investigate the processes that determine it to be what it is, and therefore what it expresses. Unlike Spinoza, Leibniz believes that determinations of nature and of human culture cannot be understood purely in terms of logic. The conditioning of S by P is not merely logical or even causal, but also historical, because the created world is the result of the intelligible and moral choice of a rational and benevolent God, the progressive expression of sentient and self-conscious, perceptive and apperceptive, creatures. Progress must be temporal; indeed, it must take place in history as well as in time. Explanation requires narrative as well as argument.

There remains, however, a tension between Leibniz's insistence on the importance of development in both the natural world and human culture, and his tendency to read temporal succession back to causal consequence and thence to logical consequence, which works against the reality of time. The latter tendency is the result of his debt to Spinoza and his love of logic and mathematics; the former stems from his interest in British empiricism and the Royal Society, and his own inexhaustible curiosity: it made him the harbinger of Goethe and Darwin, and indeed of modern scientific cosmology. For Leibniz, natural science includes natural history, but it also includes mathematical physics. Another tension exists within Leibniz's evocation of narrative: for Leibniz, time is the expression in the created world of the logical incompatibility of concepts (possibles) as space is the expression of their logical compossibility; the conditions are prior to the conditioned, and what we choose excludes the unactualized possibles we did not choose. Thus time is not illusory or indifferent, as it is for Spinoza, but rather an aspect of the orderliness and morality of the created world. However, Leibniz sometimes writes as if the time of the created world will come to an end, following the *Book of Revelations*, and sometimes as if it will go on forever, in this best of all possible worlds which is continually perfecting itself. History requires narrative, the recounting of the free acts of human beings which always include reference to the actions that were not chosen, the unactualized possibles that frame and give meaning to realized action. So Leibniz hovers between the closed moral order of *Revelations* and Dante's *Divine Comedy*, where all the stories are told in retrospect, and the open moral order of lived history.

4 Leibniz as Historian

Leibniz produced a wide range of historical writings. Two historical works that he wrote as the librarian of Hanover during the 1690s led to his commission to write a history of the House of Brunswick, which he extended to include an account of the western empire from the time of Charlemagne to the end of the Saxon imperial line. This history was not published until 1843 (two thousand pages in three volumes!), but Leibniz did publish two collections of sources, the *Accessiones Historicae* in 1697 and *Scriptores Brunsvicensia illustrantes* in 1707 and 1711, as well as many shorter works. Lessing, Kant and Herder were directly inspired by Leibniz's metaphysics and historical writings; one might observe that not only did Leibniz inspire the Enlightenment, but the Romantic turn against it, with his notion of the monad and the dynamic development of the created world. In his two-volume work *Die Entstehung des Historismus*, Friedrich Meinecke traced 20th c. historicism back from Herder to Leibniz, and Ernst Troeltsch, somewhat tendentiously, argued that Romanticism sprang from the mysticism of Eckhart and the philosophy of Leibniz (see Meinecke 1936; Troeltsch 1934).

The French historian of philosophy, Louis Davillé, traced Leibniz's accomplishments and influence as an historian directly to his metaphysics in *Leibniz historien*. At the beginning of Chap. 6, "La philosophie de l'histoire," Davillé writes, "From the metaphysical point of view, Leibniz, contemplating together the diversity and uniformity of things and beings, also follows two opposed principles, recognized earlier by scholastic philosophers, the principle of individuation and the principle of analogy, which he expresses by two phrases, in French: "l'individualité enveloppe l'infini" and "c'est tout comme ici." But this is only an appearance. Always seeking to reconcile opposites, he unites these two points of view in "la conception d'un développement à la fois spontané et régulier des êtres," through the contemplation of the universal harmony, the principle of things persisting in diversity balanced by identity. This powerful and original synthesis he calls the law of continuity… The notion of continuity plays a leading role in Leibniz's philosophy, differentiating it sharply from that of Descartes. One might call the law of continuity the 'general method' of Leibniz, and this expression doesn't seem to be an exaggeration" (Davillé 1909; my translation).

Davillé notes three formulations of the principle of continuity: (1) Time and space are divisible to infinity. (2) The order of causes or starting points is expressed in their consequences, and vice versa. This principle, 'of harmony,' is a corollary of the Principle of Sufficient Reason. It has as a corollary the principle of induction, that the whole cause can always be retrieved from the effect; the principle of differentials; and the principle of analogy. (3) Change never occurs in jumps, but always by degrees. Leibniz also calls this the principle of transition; like the principle of the identity of indiscernibles, Leibniz deduces it from the Principle of Sufficient Reason. The principle of continuity, taken as a principle governing *history*, corresponds to a conception of historical evolution, slow and successive change due to a natural and immanent cause.

The principle of continuity, Davillé observes, has obvious applications in Leibniz's mathematical and scientific research, and he clearly makes use of it in natural history (geology, botany, zoology) and human history. (The *Protogaea*, central to the development of the earth sciences in the 18th century, was recently edited and translated into English by Claudine Cohen and André Wakefield) (Leibniz 2008). In Leibniz's monadology, there is both an internal and external continuity: each monad contains within itself the series of its own development; and it expresses the world, made up of other, equally expressive monads, whose points of view differ from those of their neighbors "par des transitions insensibles." Thus the present is always pregnant with the future, and wears the traces of the past, for those who know how to read them; and the least event reverberates throughout the world. Leibniz as historian is very sensitive to the ways in which the past prolongs itself, expressing itself in the present. Inspired by the principle of continuity, Albert Sorel wrote *L'Europe et la révolution française*, which shows in eight volumes that the French Revolution was not a revolution—a discontinuity in history—after all (Sorel 1885–1904).

Not only does Leibniz's principle indicate that all historical development is an evolution, reminding us of the continuities of geography, of chronology, of genealogy; it also confers special importance on the details of culture, Davillé argues, for even the smallest thing is expressive of the whole. Thus Leibniz is always preoccupied with origins, pursues hidden causes and distant consequences, loves to digress, and explores analogy wherever he can. Thus too Leibniz always studies the languages of the cultures he studies, tracing etymologies, borrowings, and branchings on the great tree of Adam's language. Overall, the principle of continuity acts as a regulative, chastening guide, to keep historical hypotheses from veering off into improbability, and imprints on Leibniz's historical works their defining characteristics: determinism (but not fatalism), optimism, and the idea of progress.

In the past half-century, we seem to have lost sight of the importance of Leibniz as a model for philosophy of history. As Ursula Goldenbaum argues in her essay "'Peter denies'—The impact of Leibniz' Concept of Time on his Conception of History," while Hegel was interested in a teleological philosophy of history as a substitute for *spiritual history*, Leibniz was looking for a *science of history* with new methods for closing the gap between theoretical knowledge and empirical knowledge (Goldenbaum 2012). However, the temperate middle ground between glacial tradition and the fires of revolution, which Leibniz analyzed so well, has recently been re-identified by Penelope Corfield in her study *Time and the Shape of History*. Corfield identifies three central and interlocking dimensions of history in time: continuity, gradual change, and abrupt or revolutionary change. In her chapter devoted to the middle term, gradual change, she writes, "In historical application, it is helpful to distinguish between, on the one hand, slow, adaptational 'mico-change,' summarized as evolution or gradualism and, on the other hand, rapid, drastic and fundamental 'macro-change,' summarized as revolution or structural discontinuity" (Corfield 2007: 57). Leibniz's metaphysics of time invites us back to the study of micro-change, and indeed suggests that it provides the best model for historians, since cultural stasis always contains the seeds of change, and revolutions are never as revolutionary as they claim to be.

5 Leibniz as Physicist

Leibniz's definition of time as the expression (in the world of *phenomena bene fundata*) of relations of mutual incompatibility among things entails a methodological directive to the historian, to make history scientific. For Leibniz as a physicist, it led to his celebrated quarrel at the end of his life with Isaac Newton via Samuel Clarke; whereas Newton claimed that space and time are absolute and established prior to the things of the created world, for Leibniz space and time are relative to the created world, a plenum of monads. As we have seen, relativism creates a metaphysical problem, or tension, within his system, because these relations of incompatibility must not themselves be temporal, lest the definition of time be circular. Leibniz sometimes argues that these relations of incompatibility must be causal: the asymmetry of causal relations—a cause causes an effect, but an effect does not cause a cause—underlies the asymmetry of their expression as temporal relations of earlier and later. And when he speculates about possible worlds, before the creation of this world or in reference to worlds that will never be created, the asymmetry of causal relations is further referred to the asymmetry of premise and conclusion (premises support a conclusion, but a conclusion does not support a premise) or of terms in a series ($n + 1$ always comes after n); and this seems to be an echo of Spinoza, for whom time was not metaphysically real.

Thus Leibniz's speculations about the time of the cosmos proceed in a mathematical way, following the lead of Newton and employing his own supple notation for differential equations, which will become the idiom in which mechanics is developed in the 18th century. But they also proceed in a more speculative way in his metaphysical writings, like the *Discourse on Metaphysics*, where cosmic time proves both historical and moral. A few earlier manuscripts lead up to Leibniz's *Tentamen de Motuum Coelestium Causis* (1689) (Leibniz 1962, VI: 144–186). There he responds directly to Newton's *Philosophiae Naturalis Principia Mathematica* (1687). In *De Motu Gravium vel Levium Projectorum* (1688), Leibniz begins with the discovery that Newtonian curves described by rectilinear uniform motion compounded by the action of gravity can be replaced by curves described by a body pushed by a vortex rotating with a velocity inversely proportional to the distance from the center (see Bertoloni Meli 1993: 276–304).

This result allows Leibniz to claim that he can explain planetary motion just as well by appeal to the motion of vortices as by appeal to inertial motion compounded with motion caused by central forces. Newton's way of composing the trajectory is not pursued: he calculates the deviation from the tangent to the curve, the inertial path the body would have pursued if not deflected by an impulse of force. Leibniz however expresses the situation with a single differential equation, by calculating the variation of the distance from the center, comparing the distances at different times by a rotation of the radius. Leibniz writes, "Hence it is clear that if all endeavors from gravity m are added together, and from this all centrifugal endeavors k are detracted, one will have the impetus of descent ${}_1G_1L$, namely, $dr = \int m - \int k$, that is, $ddr = m - k$." Leibniz also offers an original derivation of Newton's Proposition XI (Book I) in terms of a differential equation which he solves by substitutions

$$\left(r = \left(v^2 + a^2\right)/a \text{ as well as } v = \sqrt{ar - aa}\right)$$

and separation of variables, which result in the expressions

$$dt = dr \cdot r/\sqrt{ar - aa} \text{ or } t = 2v + \int a^2 dv/v.$$

His goal is to arrive at a relation between time and distance satisfying Kepler's laws, but due to errors in calculation (the integration problem is extremely difficult), he fails to achieve it in this manuscript (see Bertoloni Meli 1993; Aiton 1985: Chap. 6).

In the *Tentamen*, section 19, Leibniz proves more successfully that "If a moving body having gravity, or which is drawn to some center, such as we suppose a planet is with respect to the Sun, is carried in an ellipse (or another conic section) with a harmonic circulation, and the centre both of attraction and of circulation is at the focus of the ellipse, then the attractions or solicitations of gravity will be directly as the squares of the circulations, or inversely as the squares of the radii or distances from the focus," by "a not inelegant specimen of our differential calculus or analysis of infinites." His reasoning results in the differential equation, $ddr = (bbaa\theta\theta - 2aaqr\theta\theta)/bbr^3$ from which Leibniz elicits $2a\theta\theta/rr$ as the 'solicitation of gravity,' which is thus "inversely as the squares of the radii." In these formalizations, time becomes merely the parameter t, a feature of the dynamical system, without the independent definition it receives in Newton's formalization, where it is represented by the virtual line of inertial motion (Leibniz 1962, VI: 156–157) (Fig. 7).

The debate between Leibniz and Newton (via his representative and friend, Samuel Clarke, in the Leibniz-Clarke correspondence) is first and foremost over the philosophical status of space and time: are space and time dependent upon, or independent of, the things and events that occur in them? (Leibniz 1978, VII: 352–440). Leibniz's plenum of monads, on the phenomenal level, conveniently 'expressed' a Euclidean spatial structure and a unidirectional, uniform temporal flow. Newton asserted by fiat that the absolute structure of space is Euclidean (with Cartesian origin and axes thrown in for good measure) and that time "flows equably without relation to anything external." Both seemed to assume, as we would

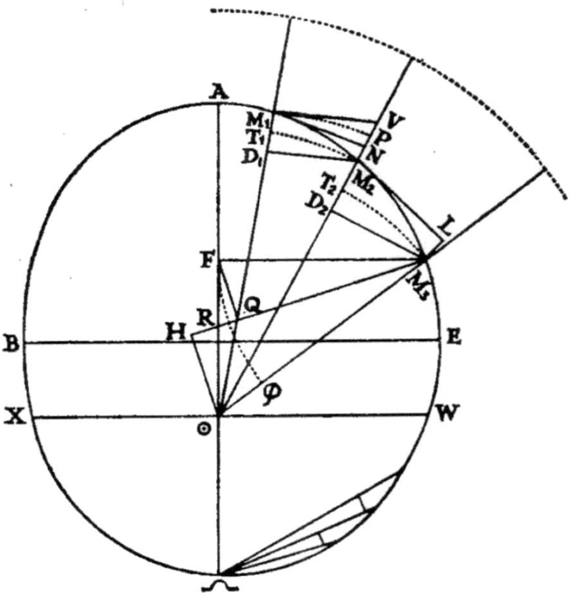

Fig. 7 The Analysis of Orbital Motion in the *Tentamen*

formulate it now, that the time-line is orthogonal to the three dimensions of space. However, because Leibniz—like Descartes—defines spatial and temporal position only in relation to other bodies, he cannot distinguish properly between inertial motion and accelerated motion; but the concept of inertial motion is central to his mechanics. And because Newton defines absolute space and time so strongly, motion itself becomes absolute, and he loses the equivalence of rest and inertial motion, which is central to his mechanics.

Both Leibniz and Newton fail to see that inertial motion is a term required to play two distinct roles. Each thinker exploits its ambiguity in order to introduce the aspect of time and space that he has neglected. In Cartesian and Leibnizian mechanics, a particle in inertial motion can never occur because the world is a plenum, and all particles are continually jostled into curved trajectories; and moreover, a Leibnizian cannot distinguish inertial motion from other curved (and thus accelerated) trajectories. However, it must remain as a stand-in for the Euclidean line and the "equable flow" of time that Newton establishes by fiat; Leibniz cannot do without absolute time, and inertial motion, by its ambiguity, provides it for him. In Newtonian mechanics, a particle in inertial motion *can* occur, in case God decides to put a single particle into the great framework of space and time: it will thus travel through the void in a straight line at a constant velocity, un-acted upon by any external forces; and moreover, it will have an absolute direction and velocity assigned to it, with respect to the immoveable, a priori, "common center of the world." However, it must also remain as the exemplum of the spatio-temporal referent, whose inherent and virtual inertial motion is constantly deflected by external forces, and in this role it must be equivalent to rest—inertial motion can have no more effect on a physical system than rest. Newton cannot do without the stipulation of time by a moving thing, and inertial motion, by its ambiguity, provides this for him (Grosholz 2011a).

Newton focuses on the geometrical representation of time as proportional to the line of inertial motion (moreover, the laws of Newtonian physics are time-reversal invariant). This approach raises the question, does the symmetry of the line show that time itself is symmetric, despite appearances? And it tends to make time seem illusory: if our best scientific representation of time contradicts empirical experience, where time is asymmetrical, perhaps time itself does not exist. Leibniz focuses on time as a feature of physical referents, the dynamical system, and thus on an empirical record of measurement. This approach raises the question, does the causal asymmetry exhibited by things (for example, the increase in entropy in the universe) show that time itself is asymmetric, or does asymmetry only happen to be a feature of appearances? And it tends to make time seem unintelligible: if the temporal features of the whole historical cosmos do not reveal anything about time, what would? In the end, I argue that a combination of both approaches is needed, to do justice to the reality and intelligibility of time, as we investigate it.

6 Leibniz as Mathematician

Leibniz believed that mathematics has a special place in the human search for wisdom, knowledge of the "most sublime principles of order and perfection," because the things of mathematics are so determinate, and exhibit their determinate inter-relations so clearly. However, the proper use of mathematics requires careful philosophical reflection. The reason why materialism has seemed attractive to serious thinkers, he argues in the *Tentamen Anagogicum* (1696), is because it lends itself well to mathematical representation, and thus to calculation and rigorous inference (Leibniz 1978, VII: 270–279). However, we should not overestimate the extent to which the material world lends itself to mathematics, for all mathematical 'models' are a finitary representation of an infinitary reality; and we should not forget that other aspects of reality also lend themselves similarly to mathematization. The materialist illusion is not only a mathematical mistake (which should be addressed by yet more mathematics) but also a metaphysical mistake. The alleged materialist universe is a mirage, for it violates the Principle of Sufficient Reason, which along with the principle of contradiction governs the created world; it is thus after all not thinkable, like the mirage of the 'greatest speed.' The world's beings are not only material, but thoroughly sentient and endowed with force or conatus, a striving for perfection; and in that striving they express their Maker, as well as the intelligibility for which mathematics is apt. Leibniz writes that "The ancients who recognized nothing in the universe but a concourse of corpuscles," as well as the modern philosophers who are inspired by them, find materialism plausible, "because they believe that they need to use only mathematical principles, without having any need either for metaphysical principles, which they treat as illusory, or for principles of the good, which they reduce to human morals; as if perfection and the good were only a particular result of our thinking and not to be found in universal nature… It is rather easy to fall into this error, especially when one's thinking stops at what imagination alone can supply, namely, at magnitudes and figures and their modifications. But when one pushes forward his inquiry after reasons, it is found that the laws of motion cannot be explained through purely geometric principles or by imagination alone" (Leibniz 1978, VII: 271).

Moreover, he adds, there is no reason to suppose that other phenomena which in that era had eluded mathematical formulation (he mentions light, weight, and elastic force) will not sooner or later prove to lie within the expressive powers of mathematics. But all such representation will be provisional, because while finitary models can express the infinitary things of nature well, they can never express them completely; and the formulation of increasingly accurate stages of representation must be governed, like nature itself, by the two great principles of contradiction and sufficient reason.

Leibniz recognizes that different sciences require different methodologies, but no matter what special features different domains exhibit, he believes that all scientific investigation must move between mathematics and metaphysics. Mechanics, in particular, is best viewed as a middle term between mathematics and metaphysics, and so too Leibniz's account of time. Of all the parameters involved in mechanics, time is the least tied to any specific content, even though it presents a determinate topic for scientific investigation. Thus a closer look at Leibniz's account of time presents an especially 'pure' version of the interaction of mathematics and philosophy in the service of progressive knowledge.

As Yvon Belaval, Gilles-Gaston Granger, François Duchesneau, and Daniel Garber have variously argued on the basis of a wide range of texts, Leibniz's novel conception of scientific method has two dimensions (Belaval 1960; Granger 1981; Duchesneau 1993; Garber 2009). His account of method is informed by that of Bacon and Descartes, but diverges from both in significant ways and combines aspects of each. He borrows from Bacon the project of collecting empirical samples from the laboratory and field, inductively, and compiling tables, taxonomies and encyclopediae, always with the expectation of discovering harmonies and analogies, deeper systematic organization in the things of nature. He borrows from Descartes (and Spinoza) the assurance that the indefinite presentations of sense can be associated with precise mathematical concepts, and thus by analogy be re-organized as ordered series, which can then be subject to deductive inference.

In the *Tentamen Anagogicum*, Leibniz mentions the use of geometry in the "analysis of the laws of nature," and goes on in that essay to develop the ideas of Fermat, Descartes, and Snell in optics using a series of geometrical diagrams, as well as the ideas of maximal and minimal quantities developed in his infinitesimal calculus. In an earlier, more general essay, "Projet d'un art d'inventer," (1686), he invokes arithmetic as a source of formulations apt for analysis considered as the art of invention, "which would have the same effect in other subject matters, like that which algebra has on arithmetic. I have even found an astonishing thing, which is that one can represent all kinds of truths and inferences by means of numbers"

(Leibniz 1961: 175). The idea is to locate nominal definitions, involving a finite number of requisites, and then reason on the basis of them: "I found that there are certain primitive terms—if not absolutely primitive then at least primitive for us—which once having been constituted, all our reasonings could be made determinate in the same way as arithmetical calculations; and even in the case of those reasonings where the data, or given conditions, don't suffice to determine the question completely, one could nevertheless determine [metaphysically] mathematically the degree of probability." The clarity and determinacy of mathematical things is crucial to this method of analysis. "The only way to improve our reasonings is to make them as salient as those of mathematicians, so that one can spot an error clearly and quickly, and when there is a dispute, one need only say: let us compute, without further ado, to see who is right" (Leibniz 1961: 176, my translation).

Early modern mechanics begins by exploiting an already existing trove of empirical records, the precise tables left by centuries of astronomers tracking the movements of the moon, the planets, certain stars and the named constellations which culminate in the careful data of Tycho Brahe, so important to Kepler, and which are soon thereafter improved by the measurements of astronomers equipped with telescopes. Happily for human science, the solar system is both an exemplary mechanical system (just a few moving parts, isolated, and so almost closed despite the occasional comet) and a very precise clock; so its study richly repaid the efforts of early modern physicists.

How shall these two occupations, empirical compilation and theoretical analysis, be combined? Leibniz calls on metaphysics, in particular the Principle of Sufficient Reason in the guise of the principle of continuity, to regulate a science that must be (due to the infinite complexity of individual substances) both empirical and rationalist. The correlation of precise empirical description with the abstract conception of science *more geometrico* is guaranteed by the thoroughgoing intelligibility and perfection of the created world, and encourages us to work out our sciences through successive stages, moving back and forth between a concrete taxonomy and abstract systematization. Empirical research furnishes nominalist definitions—finite lists of requisites for the thing defined—which can set up the possibility of provisionally correct deductions, though every such definition due to its finitude can be corrected and amplified; mathematics provides the rule of the series.

This model of scientific inquiry accords very well with Leibniz's own investigations into mechanics and planetary motion, and so too his mathematical-metaphysical account of time. Given the subtlety of his conception of method, I will argue that his account of time is deeper and more multivalent than that of Newton, which explains why it has proved to be more suggestive for physicists in succeeding eras and especially during the last century. Here we will take a step back, and look at Descartes' account of time, against which Newton reacts.

7 The Dispute Between Leibniz and Newton

Descartes' definition of motion in the *Principles of Philosophy* is "the transfer of one piece of matter, or one body, from the vicinity of those bodies which are in immediate contact with it, and which are regarded as being at rest, to the vicinity of other bodies" (Descartes 1964–1974, VIII: 53). Thus motion and rest can be interpreted only as a difference in velocity or acceleration established with respect to a reference frame of other bodies; no absolute determination of motion or rest is possible. This definition of motion and rest is so radically relativistic that, strictly speaking, the Cartesian observer, by choosing different reference frames, may not only shift from judging that a given particle is at rest to judging that it is in inertial motion (rectilinear motion at a constant speed), but also to judging that its trajectory should be considered accelerated (and perhaps curvilinear). Descartes himself never seems to have considered this consequence of his relativism, nor its inconsistency with his invocation of inertial motion in the first two rules of motion given at the beginning of the *Principles*. Perhaps the inconsistency escaped his notice because in his mechanics there is no accelerated motion: the inherent motion of corpuscles is rectilinear and constant in speed (that is, inertial) and the transfer of momenta (defined for each contributing corpuscle as bulk times constant speed) in a collision is instantaneous. His mechanics is thus undynamical and atemporal; its laws are not only time-reversal invariant, they do not involve time as an independent variable: nothing in Descartes' mechanics varies continuously with respect to time.

Newton, however, saw and criticized this outcome, precisely because it entails that Descartes is not entitled to his own definition of inertial motion. In *De Gravitatione* (unpublished in his lifetime) he argues that since in Cartesian vortex mechanics all bodies are constantly shifting their relative positions with time, "Cartesian motion is not motion, for it has not velocity, nor definition, and there is no space or distance traversed by it. So it is necessary that the definition of places, and hence of local motion, be referred to some motionless thing such as extension alone or space in so far as it is seen to be truly distinct from bodies" (Newton 1962: 131). That is, Descartes cannot give empirical procedures in his mechanics that allow him to distinguish inertial motion from accelerated motion.

Newton responds with his well-known thought experiment about the revolving bucket, arguing that the presence of forces is the sign of true (accelerated) motion; forces are real and measurable. But he goes beyond that claim: in Book III of the *Principia*, he writes,

> Hypothesis I: The center of the system of the world is at rest.
>
> Proposition 11, Theorem 11: The common center of gravity of the earth, the sun, and all the planets is at rest (Newton 1972, II: 816).

Taken together, these claims offer an absolutist conception of space that makes not only accelerated motion, but even uniform motion, definable with respect to a Euclidean space that has been provided with a centre and axes. By countering so strongly Descartes' relativism and subsequent loss of the distinction between inertial motion and accelerated (straight or curvilinear) motion, Newton has sacrificed the equivalence of inertial reference frames and thus his own first law, as we noted above. He has also postulated a spatio-temporal structure that cannot be empirically verified, a set of Cartesian coordinates for the Euclidean space of his planetary mechanics, which violates his methodological principle of not invoking merely metaphysical hypotheses.

Newton is not entitled to the equivalence of rest and inertial motion, which is just as essential to his system as Descartes' concept of inertial motion is to his system (Grosholz 2011b). Leibniz acknowledged but was not troubled by the consequences of Descartes' relativism, and extended it to time. Thus in a commentary on the *Principles*, "Critical Thoughts on the General Part of the Principles of Descartes," (unpublished in his lifetime), Leibniz writes about *Principles* II, Articles 25 and 26:

> If motion is nothing but the change of contact or of immediate vicinity, it follows that we can never define which thing is moved. For just as the same phenomena may be interpreted by different hypotheses in astronomy, so it will always be possible to attribute the real motion to either one or the other of the two bodies which change their mutual vicinity or position. Hence, since one of them is arbitrarily chosen to be at rest or moving at a given rate in a given line, we may define geometrically what motion or rest is to be ascribed to the other, so as to produce the given phenomena. Hence if there is nothing more in motion than this reciprocal change, it follows that there is no reason in nature to ascribe motion to one thing rather than to others. The consequence of this will be that there is no real motion (Leibniz 1978, IV: 369).

This is just what Newton says! But for Leibniz, it is not a problem, certainly not a problem to be banished by postulating absolute space and time as the arena for motion. Rather, he makes the following claim in "Animadversiones in partem generalem Principiorum Cartesianorum": "Thus, in order to say that something is moving, we will require not only that it change its position with respect to other things but also that there be within itself a cause of change, a force, an action" (Leibniz 1978, IV: 369). Newton proposes that whenever acceleration occurs, it is due to the action of forces; Leibniz proposes that whenever any motion occurs, it is due to the action of forces. This doesn't mean that he has reverted to Aristotelianism, but is instead an expression of his pan-animism. What Leibniz means by force is not Newtonian force, but something more like energy, internal to the body. Leibniz believes that no body is ever truly at rest, for all bodies are ensouled: motion thus becomes an expression of *conatus*, as individual substances jostle each other for a place within the Cartesian plenum at all times (Leibniz 1978, IV: 350–392).

In this picture of the universe, we see the Principle of Sufficient Reason at work, fashioning Lebniz's mechanics along with mathematics. The universe must be a plenum, and the individual substances in that plenum are jostling each other in an effort to attain perfection: everything strives. Indeed for Leibniz even unactualized possibles strive: essences strive for existence. In the realm of ideas, this striving sorts ideas out into an infinity of possible worlds, and (with the beneficent cooperation of God) precipitates one world into creation; in the created world, it induces vortical motion in the plenum as well as temporality. Time is the expression of the incompatibility of things; because creation involves plurality, mentality, and mutual limitation, all things are active, passive and intentional. This is the best of all possible worlds because it is continually becoming more perfect, on into the infinite open future: creation is a continuous temporal process. In the law of the series, the independent variable is always time. Thus matter is not merely extended, but involves resistance and action; and it develops: Leibniz's science will also be a natural history. Thus his ideas stand at the origin not only of modern biology, but also of modern cosmology.

Having invented a supple and powerful notation for his version of the infinitesimal calculus during his sojourn in Paris (1672–76), Leibniz proceeded to work out a theory and practice of differential equations, in which the dependence of different forms of accelerated motion on time could be clearly expressed by the term 'dt'. One application of this method was to planetary motion. While in Vienna on his way to Rome in 1688, Leibniz read Newton's *Principia*, took extensive notes and then wrote a series of papers that culminated in the *Tentamen de Motuum Coelestium Causis* (*Acta Eruditorum*, Feb. 1689), where he proposed differential equations that would characterize planetary motion. Leibniz combined Cartesian vortex theory with Newton's reformulation of Kepler's laws, locating the planets in 'fluid orbs' rather than empty space, in order to derive the laws governing central forces while avoiding the problem of action at a distance. Whereas Newton calculates the deviation from the tangent to the curve, as we saw above, Leibniz expresses the situation with a single differential equation, by calculating the variation of the distance from the center, comparing the distances at different times by a rotation of the radius. The upshot of his calculation, is that the effect of gravity can be written as

$$\left[\left(2h^2\right) / \left(ar^2\right)\right]\, dt^2$$

so that the 'solicitation of gravity' (conceptualized in Cartesian terms as the action of a vortex) is inversely proportional to the square of the distance, which was of course the result Leibniz was trying to reproduce.

For Leibniz, space is the expression in the created world of the logical order of compossibility among individual substances, and time is the logical order of incompatibility among individual substances (he asserts this in, for example, his Letter to de Volder of June 20, 1703) (Leibniz 1978, II: 248–253). Thus, space and time only come into being with the creation of this material universe, the best of all possible worlds, and have only a secondary ontological status, because they are constituted as relational structures of the things with primary ontological status, individual substances. This is the basis of Leibniz's relationalism; but we must recall that his relationalism is deployed on the basis of a method which, as we have seen, is two-tiered, both mathematical (seeking a precise mathematical correlate for the law of the series) and metaphysical while at the same time empirical (examining and tabulating evidence in an ongoing search for the systematic organization of things). The true scientist will find ways to put the mutual adjustment of nominalistic form with the investigation of the infinitely complex, infinitely ordered world of individual substances, in the service of the progress of knowledge; this process requires both mathematics and metaphysics.

To correlate time with precise mathematical concepts, Leibniz chooses as the correct representation the straight Euclidean line, endowed with directionality by Descartes' analytic geometry, which assigns positive and negative numbers—real numbers we would say—to the line. In some texts, it appears that Leibniz holds time to be a half-line, given what he writes to Clarke in the fifth letter of the Leibniz-Clarke correspondence (Leibniz 1978, VII: 389–420). Since this is the best of all possible worlds, created by God, the universe must constantly increase in perfection, and so has a temporal beginning point but no end. Thus it is metaphysically important that the number-line is both geometrical and arithmetical. As arithmetical, it expresses the fact that time is asymmetric; time may be counted out in units, like seconds or years, and the numbers increase in a unidirectional order without bound to infinity. The asymmetry of time follows from the metaphysical ground that everything strives. As geometrical, the number-line expresses the fact that time is a continuum; units of time like seconds are not atoms, but conventionally established, constant measures of time, as the inch is a measure of continuous length. An instant is only the marker of a boundary of a stretch of time, not what time is composed of; we misunderstand what an instant is, Leibniz observes, if we conceive of it as an atom of time. Time must be both measured and counted.

This duality of time is not however without conundrums. Analysis in arithmetic leads us to the unit; but in geometry it leads us to the point. Whole numbers are composed of units, but lines are bounded by points, not composed of them; Cartesian reductionism is useful as an approach to arithmetic, but not to geometry. In a letter to Louis Bourguet, composed just before the correspondence with Clarke, in August 1715, Leibniz writes,

> As for the nature of succession, where you seem to hold that we must think of a first, fundamental instant, just as unity is the foundation of numbers and the point is the foundation of extension, I could reply to this that the instant is indeed the foundation of time but that since there is no one point whatsoever in nature which is fundamental with respect to all other points and which is therefore the seat of God, so to speak, I likewise see no necessity whatever of conceiving a primary instant. I admit, however, that there is this difference between instants and points—one point of the universe has no advantage of priority over another, while a preceding instant always has the advantage of priority, not merely in time but in nature, over following instants. But this does not make it necessary for there to be a first instant. There is involved here the difference between the analysis of necessities and the analysis of contingents. The analysis of necessities, which is that of essences, proceeds *from the posterior by nature to the prior by nature*, and it is in this sense that numbers are analyzed into unities. But in contingents or existents, this analysis *from the posterior by nature to the prior by nature* proceeds to infinity without ever being reduced to primitive elements. Thus the analogy of numbers to instants does not at all apply here. It is true that the concept of number is finally resolvable into the concept of unity, which is not further analyzable and can be considered the primitive number. But it does not follow that the concepts of different instants can be resolved finally into a primitive instant (Leibniz 1978, III: 580–583, my translation).

The analysis of time requires the scientist to proceed both by the analysis of contingents, using the line whose continuity is the best expression mathematics provides for infinite complexity; and by the analysis of necessities, using the natural numbers whose linear ordering and asymmetry is the best mathematical expression of irrevocability. Leibniz goes on to observe that the use of mathematics does not solve the metaphysical question whether time has a beginning, which leads one to suppose that more metaphysics and more empirical research are required. He writes:

> Yet I do not venture to deny that there may be a first instant. Two hypotheses can be formed—one that nature is always equally perfect, the other that it always increases in perfection. If it is always equally perfect, though in variable ways, it is more probable that it had no beginning. But if it always increases in perfection (assuming that it is impossible to give its whole perfection at once), there would still be two ways of explaining the matter, namely, by the ordinates of the hyperbola B or by that of the triangle C (Leibniz 1978, III: 580–583, my translation).

Thus despite what he would shortly write to Clarke, he was perhaps not convinced that time has a beginning:

> According to the hypothesis of the hyperbola, there would be no beginning, and the instants or states of the world would have been increasing in perfection from all eternity. But, according to the hypothesis of the triangle, there would have been a beginning. The hypothesis of equal perfection would be that of rectangle A. I do not yet see any way of demonstrating by pure reason which of these we should choose. But though the state of the world could never be absolutely perfect at any particular instant whatever according to the hypothesis of increase, nevertheless the whole actual sequence would always be the most perfect of all possible sequences, because God always chooses the best possible (Leibniz 1978, III: 580–583, my translation).

In any case, Leibniz's conception of method requires that time be investigated not solely by pure reason or pure mathematics, which he admits here to being inconclusive; time must also be investigated empirically. It must be considered as the relational structure of the individual substances that exist, insofar as they are not logically compatible with each other. This means that we may have to revisit the formal structures we have just been discussing, in light of what we discover about the physical universe. The Principle of Sufficient Reason governs the created world; not only does it entail that everything is determinate and intelligible (which for Leibniz means, thinkable), it also entails that everything strives for perfection. Thus the essences that are ideas in the mind of God strive for existence, but only those that constitute this best of all possible worlds succeed; and in the created world, the essences continue to jostle each other, to interfere with each other, as they all strive. This dynamic quality of ideas produces time, as their harmonies produce space; creation entails plurality and mutual limitation, activity and passivity. And the time that is produced is asymmetrical, as creation tends towards greater perfection, a harmonious dissension among the sentient, active individual substances.

What Leibniz heralds is the now received belief that matter is not passive and inert, or dead: even a molecule is mobile, active, forceful, and sensitive. As he writes in the *Monadology*, sec. 66–69:

66. (…) there is a world of creatures, of living beings, of animals, of entelechies, of souls in the least part of matter.

67. Each portion of matter can be conceived as a garden full of plants, and as a pond full of fish. But each branch of a plant, each limb of an animal, each drop of its humors, is still another such garden or pond.

68. And although the earth and air lying between the garden plants, or the water lying between the fish of the pond, are neither plant nor fish, they contain yet more of them, though of a subtleness imperceptible to us, most often.

69. Thus there is nothing fallow, sterile, or dead in the universe, no chaos and no confusion except in appearance (…) (Leibniz 1978, VI: 618–619).

8 Coda

Leibniz understands that productive scientific and mathematical discourse must carry out distinct tasks in tandem: a more abstract search for conditions of intelligibility or solvability, and a more concrete strategy for achieving successful reference. While deductive argument is important (since its forms guarantee the transmission of truth from premises to conclusion) as a guide to effective mathematical and scientific reasoning, it does not exhaust method, for Leibniz. As we have seen, Leibnizian method has two dimensions, empirical and rational, and both require analysis, whose logical structure includes abduction, induction, and reasoning by analogy, as well as deduction. Moreover, analysis, the search for conditions of intelligibility, is more than logic; it is a compendium of research and problem-solving procedures, which vary among investigations of different kinds of things. Such representational combination and multivocality is just what we find in Leibniz's most important pronouncements on the nature of time.

References

Augustine of Hippo. (1961). *Confessions*. R. S. Pine-Coffin (Tr.). New York: Viking Penguin.

Aiton, E. (1985). *Leibniz: A biography*. Bristol: Adam Hilger.

Belaval, Y. (1960). *Leibniz critique de Descartes*. Paris: Gallimard.

Bertoloni Meli, D. (1993). *Equivalence and priority, Newton versus Leibniz*. Oxford: Oxford University Press.

Corfield, P. (2007). *Time and the shape of history*. New Haven, Ct.: Yale University Press.

Davillé, L. (1909). *Leibniz historien. Essai sur l'activité et la méthode historique de Leibniz*. Paris: Félix Alcan.

De Gandt, F. (1995). *Force and geometry in Newton's Principia*. Princeton: Princeton University Press.

Descartes, R. (1964–1974). *Oeuvres de Descartes*. C. Adam and P. Tannery (Eds.) 11 Volumes. Paris: Vrin.

Duchesneau, F. (1993). *Leibniz et la méthode de la science*. Paris: Presses Universitaires de France.
Galilei, G. (1914/1954). *Dialogues concerning two new sciences*. H. Crew & A. de Salvio (Tr.). New York: Dover.
Garber, D. (2009). *Leibniz: body, substance, monad*. New York: Oxford University Press.
Goldenbaum, U. (2012). "Peter Denies"—The impact of Leibniz's conception of time on his conception of history. *Revue Roumaine de Philosophie, 56*(1), 79–88.
Granger, G.-G. (1981). Philosophie et mathématique leibniziennes. *In Revue de Métaphysique et de Morale, 86*, 1–37.
Grosholz, E. (2007). *Representation and productive ambiguity in mathematics and the sciences*. Oxford: Oxford University Press.
Grosholz, E. (2011a). Reference and analysis: The representation of time in Galileo, Newton, and Leibniz. *Journal of the History of Ideas,* 72/3, 333–350.
Grosholz, E. (2011b). Space and time. In D. Clarke & C. Wilson (Ed.), *The Oxford handbook of philosophy in early modern Europe*. Oxford: Oxford University Press. 51–70.
Koyré, A. (1939). *Etudes galiléennes*. Paris: Hermann.
Leibniz, G. W. (1961). *Opuscules et fragments inédit de Leibniz*. L. Couturat (Ed.) Hildesheim: Georg Olms.
Leibniz, G. W. (1962). *Mathematischen Schriften*. C. I. Gerhardt (Ed.), 7 Volumes. Hildesheim: Georg Olms.
Leibniz, G. W. (1978). *Philosophischen Schriften*. C. I. Gerhardt (Ed.), 7 Volumes. Hildesheim: Georg Olms.
Leibniz, G. W. (2008). *Protogaea*. C. Cohen & A. Wakefield (Ed.). Chicago: University of Chicago Press.
Meinecke, F. (1936). *Die Entstehung des Historismus*. Munich: R. Oldenbourg.
Newton, I. (1934). *Mathematical principles of natural philosophy and his system of the world*. In A. Motte (Tr.) and F. Cajori (Ed.). Berkeley: University of California Press.
Newton, I. (1962). De gravitatione et aequipondio fluidorum. In *Unpublished scientific papers of Isaac Newton*. A. R. Hall and M. B. Hall (Ed.). Cambridge: Cambridge University Press, 89–156.
Newton, I. (1972). *The* Principia, *mathematical principles of natural philosophy* [3rd Edition, 1726.] I. B. Cohen and A. Whitman (Tr.) 2 Volumes. Berkeley: University of California Press.
Sorel, A. (1885–1904). *L'Europe et la révolution française*. 8 vols. Paris: Plon, Nourrit et Cie.
Troeltsch, E. (1934). The ideas of natural law and humanity in world politics. In O. Gierke (Ed.) E. Barker (Tr.) *Natural law and the theory of society*. Cambridge: Cambridge University Press: Appendix I.
Van Fraassen, B. (1989/2003). *Laws and symmetry*. Oxford: Oxford University Press.

Chapter 7
The Representation of Time from 1700 to the Present

To probe the limits of Leibniz's relationalism in order to lead into a discussion of the treatment of time in modern physics in the wake of Newton and Leibniz's celebrated dispute, I leave the path of textual analysis for a couple of pages, and venture into the forest of thought experiments. Inspired by 20th c. speculation, I propose that we try out Leibnizian relationalism on models of the universe very different from that which he entertained, and see what becomes of the account of time. First, let us suppose that nothing exists except a single particle. Then there is no time, because time is the expression of relations of incompatibility among things and one thing is clearly compatible with itself.

Suppose next that nothing exists except a perfect harmonic oscillator, which moves through a certain series of configurations only to return to exactly the same configuration in which it began. The motion of the harmonic oscillator, with one causal state giving rise to another, expresses time, but is the time it expresses finite or infinite? Since its beginning and end state are identical, it seems as if we should identify the times they express; then time would be finite. The local 'befores' and 'afters' would have no global significance; the asymmetry of cause and effect along the way would be absorbed into a larger symmetry, because every effect would ultimately be the cause of the cause... of the cause of its cause. Thus the local incompatibility of before and after would be absorbed into a global compatibility; but then we must wonder whether this finite time is really temporal at all. It seems that in this picture duration both does and does not occur.

Moreover, the picture seems to contradict the supposition that what exists is a perfect harmonic oscillator, for there is no oscillation. The concept of oscillation involves the notion of repetition, which in turn requires a linear ordering of time, so that when a particular configuration recurs, that is when it occurs again, the first occurrence is earlier than the later one, but the later one is not earlier than the first. We can imagine that the same configuration recurs at a later moment of time; but it is incoherent to suppose that *the selfsame moment of time* recurs at *another moment of time*, for those two moments of time must then be both identified with, and distinguished from, each other. As Leibniz often observes, contradiction makes

E.R. Grosholz, *Starry Reckoning: Reference and Analysis in Mathematics and Cosmology*, Studies in Applied Philosophy, Epistemology and Rational Ethics 30, DOI 10.1007/978-3-319-46690-3_7

alleged ideas vanish into nothingness; the relationalist idea of an isolated harmonic oscillator is a mirage, and so is the idea of a moment of time recurring.

So we would have to admit that the time that frames the harmonic oscillator is ongoing, linear and infinite, and so must be constituted by something beyond the relations that hold among the moving parts of the harmonic oscillator; but this goes against Leibnizian relationalism. To avoid this problem, Leibniz must completely fill up his cosmos with things and events that never repeat, on pain of incoherence. Such a cosmos is precisely what his metaphysics provides, chosen by God according to the Principle of Plenitude, the Principle of Perfection, the Principle of Sufficient Reason, and the Principle of Contradiction. Moreover, since all of his monads are body-souls, everything that exists is provided with a developed or rudimentary intentionality, that drives it forward in time. The strong asymmetry observed in the organic, sentient world is guaranteed for everything that exists. In Leibniz's cosmos, everything is alive and everything strives. The dispute with Clarke shows that Leibniz's cosmos must be a plenum, for otherwise isolated things would show up in absolute space and God's choice of their location would be arbitrary; similarly, if isolated events happened in absolute time, God's choice of when they occurred would be arbitrary. So even if we imagine the ideal harmonic oscillator to express an ongoing, infinite time, perhaps by allowing the natural numbers as a condition of its intelligibility, so that each of its oscillations might thereby be distinguished by a numerical index, it would still violate the Principle of Sufficient Reason.

At this juncture in the argument, however, we might suspect that Leibniz has not discovered the infinity and uni-directionality of time in the relations among things, but merely construed the relations among things so that the time they express will turn out to be appropriate, that is, infinite and uni-directional. And another suspicion may arise: Even if Leibniz is accurately describing the way things are (an organicist, animist plenum), perhaps that in itself sheds no light on time. Time itself may have no flow; and it may prove to be finite, coming to an end that no living thing (including Leibniz) foresees. If our grasp of time is merely empirical, based on temporal relations among things, maybe real time is beyond our grasp. However, for Leibniz no pursuit of truth should be merely empirical; to be a Leibnizian relationalist is not to reduce science to empiricism. Leibniz avoids this skeptical worry by trusting in the ability of metaphysical principles to regulate the interaction of empirical research and theoretical speculation in science. Informing this trust is his trust in the perfection and intelligibility of the cosmos, so that time is the expression of the infinite, harmonious incompatibility of things.

1 Penrose and Smolin in the Long Shadow of Leibniz

Variations of this problem continue to this day. Here are two of them, resurfacing in two recent books by prominent cosmologists who acknowledge the influence of Leibniz, Roger Penrose and Lee Smolin. In a recent book, *Cycles of Time: An*

Extraordinary New View of the Universe, Roger Penrose suggests a "conformal cyclic cosmology". He observes that as we move back in time, temperatures are increasingly great, so that all particles are so energetic as to be effectively massless near the Big Bang. Massless particles in relativity theory do not experience the passage of time, but they and their interactions may be supposed to satisfy "conformally invariant equations". At this point, the universe cannot build a clock and so loses track of time, while retaining its conformal geometry. Penrose then applies similar reasoning to the distant future, where the universe again forgets time in the sense that there is no way to build a clock with just conformally invariant material, and concludes that with conformal invariance both in the remote future and at the Big Bang origin, we can try to argue that the two situations are physically identical, so the remote future of one phase of the universe becomes the Big Bang of the next. This results in his conception of *conformal cyclic cosmology*. The structure of this thought experiment is not so different from the one I proposed above, concerning a universe composed of a single perfect harmonic oscillator, whose beginning state can be identified with its end state (see Penrose 2011).

Making use of the four-dimensional space-time manifold of Relativity Theory, Penrose suggests that at the two identified end and beginning states, temporality has disappeared because massless particles are atemporal, and the universe is "space-like". When temporality is not expressed by the things that exist, it fades away, so that the duration of the universe is not finite but also not infinite, as if it had the topology of an open interval. Thus temporality, fading in and then fading out, is located somewhere between the beginning and end states. However, the problems that arise with respect to the harmonic oscillator arise here as well. Even if at the beginning and end of the cosmos time has faded away, Penrose still speaks of phases of the universe, one of which is next after ours. The very notion of an event being repeated (because it is physically identical to another, in a different location of spacetime) entails a linear ordering of time; if it is not the time of the massless, temporally indifferent things of the universe at the beginning and end state(s), then it is a meta-time, and the relationalism of Leibniz and Penrose must be modified: relational time must be supplemented by absolute time. If the hypothesis of relationalism is not forgone, then we have a location in spacetime occupying another location in spacetime, which seems just as incoherent as the notion that a moment can occupy another moment. Is cosmology the mathematical description of the perennial structure of the whole universe, an extension of Newtonian-Leibnizian or rather Einsteinian mechanics, or is it a kind of natural history? I have been trying to show that we owe this modern conundrum in part to Leibniz, whose metaphysics of time introduced the theoretical possibility that the scientific treatment of the cosmos might also be historical.

Lee Smolin takes up the question of the reality of time in his recent book *Time Reborn* (Smolin 2013). He advocates a novel version of naturalism, proposing that the passage of time is real, the present moment is real and the past consists of moments that were real (but the future is open), the universe is unique, and the laws of nature evolve in time. His position is informed, he announces, by Leibniz's Principle of Sufficient Reason augmented by the claim that the universe contains all its causes, and the Principle of the Identity of Indiscernibles, and his attendant philosophy of relationalism. Here Smolin counters a number of assumptions often associated with earlier versions of naturalism: everything that exists in the natural world can be described in the language of physics, the laws of physics are invariant, and the universe really is an invariant mathematical object: the passage of time is an illusion from the point of view of the "block universe". Smolin argues that this version cannot explain why our experience is organized as a one-dimensional succession of moments, why we remember the past but not the future, and how we use the word "now". He argues further that block universe naturalism stems from an illegitimate extension of the Newtonian paradigm, to study an isolated subsystem, idealized and observed from outside; Book I of the *Principia* indicates how crucial this strategy was for launching modern physics. However, it is highly misleading when we are trying to understand the whole universe; there is no outside and there is no external clock.

Smolin thus proposes his own version of naturalism, which he believes will allow cosmology to seek better answers, both theoretically and empirically, to important questions: What are the conditions that give rise to the laws of physics? How might these laws evolve? Why does the universe display the asymmetry of time (why has it been so long out of equilibrium)? Are the laws of nature time reversal invariant or not? Is the future open? His proposal makes use of the theory of 'shape dynamics' (also explored by Julian Barbour), an interpretation of general relativity in which there is a preferred slicing of space-time, that is, a preferred choice of time coordinate that has physical meaning. This confers physical meaning on the simultaneity of distant events; but physics on these fixed slices is invariant under local changes of distance scale. The preferred slices are called constant mean curvature slices, and offer a global notion of time for the universe. Other features of his position include viewing energy and momentum as intrinsic rather than relational, considering events as unique (repeatable laws depend on coarse graining, which forgets the information that makes events unique), and adding novel ways to characterize and investigate the evolution of natural laws, a 'principle of precedence' and a 'principle of cosmic natural selection'. His speculations show (whether one tends towards Barbour's conviction that time is not real, or Smolin's conviction that it is) how crucial our understanding of the nature of time is for the development of modern physics. Towards the end of his paper, in a discussion of *qualia* or conscious experience (including my experience of a "now"), Smolin suggests that the conviction that time is real might lead one to entertain a revival of Leibniz's pan-psychism. That is, *qualia* cannot be part of the block universe and time, but if—as temporal naturalism assumes—time is real, they might be important aspects of the natural world (see also Smolin 2015).

2 The Dispute Between Leibniz and Newton Redux

As we have seen, Newton claims that space and time are given independent of anything that might be set into them. Euclidean space offers the correct structure for space, and "absolute, true and mathematical time, of itself, and from its own nature, flows equably without relation to anything external" (Newton 1934: 6). It is thus nicely represented by the inertial motion of a body, a straight line flowing equably. Newton solves Galileo's unacknowledged problem with the inertial frames by assigning the world system a center and a coordinate system, which makes all motion absolute and thus, as noted above, impugns the physical equivalence of rest and inertial motion, which Newton nonetheless invokes for his system. Leibniz, by contrast, asserts that space and time are nothing in themselves, but rather the expression of underlying causal or logical relations, holding among the things that exist. Space and time are relations of compatibility or incompatibility among things; and things are dynamical for Leibniz because for him all matter is ensouled and therefore endowed with force and a lesser or greater sentience and intentionality. For Leibniz, as for Goethe and Smolin, everything strives. Thus for Leibniz, time finds its expression in differential equations applied to mechanics; time is no more than the record of things that change with time.

Newton's absolutist metaphysics leads him to prefer the abstract, geometrical, autonomous representation of time, and Leibniz's relational metaphysics leads him to prefer the concrete representation of time that arises from the dynamical object, given in terms of differential equations. The metaphysical and mathematical preferences that divide the two men are very strong. The mathematics that Newton uses almost polemically in the *Principia* is geometry. He undermines his own development of infinitesimalistic methods; the variables used by Newton are finite, because he tries to sidestep the problem of infinitesimals by using finite ratios between vanishing or 'evanescent' quantities. And Newton's notation does not lend itself to the development of differential equations, as becomes clear in Book I, Propositions XXXIX, XXXX, and XXXXI, where he tackles and does not completely solve the problem of finding the shape of a trajectory given that the law obeyed by the central force is the inverse square law (De Gandt 1995: 244–264). By contrast, Leibniz puts his entire treatment of planetary motion into the elaboration of his differential equations, encouraged by his own lucid and suggestive notation, but loses command of the formalization because he sets the solar system into Cartesian vortices ('fluid orbs') instead of Newton's useful and simple Euclidean space.

What conclusions can we draw here? The debate between Newton and Leibniz may be viewed as inevitable, but irresolvable. The issue is not whether one should try to understand time as an abstract structure independent of the things in time, *or* as a constitutive feature of the concrete things in time. My point is that scientists must try to do both, in a discourse with at least two disparate registers that, despite the disparity and also in part because of it, still manages to make sense and to reveal important things about the world. And indeed, that is what both Newton and

Leibniz did in their scientific and mathematical practice. Every mode of representation is a distortion as well as an illumination, so we must always remain reflective about those we use. And every disparity will sooner or later produce contradictions that must be attended to, though *pari passu* the combination of disparate discourses also increases information and makes insight somehow more visible. Moreover, finally, we need to refer successfully and analyze at the same time, in order to say something true.

How do scientists manage to refer successfully and analyze deeply at the same time? What terms play the role of concrete or abstract, and how is their disparity mediated? The answers to these questions must be determined by historical study; there is no universal answer. All the same, they must still be raised by philosophy of science, which then pretty clearly can't proceed without history of science. My reading of Galileo, Newton, and Leibniz has been influenced retrospectively by recent debates in 20th c. and early 21st c. cosmology, where I discern the same tension between abstract and concrete descriptions of time, and the same kind of misunderstandings over competing metaphysical schemes (see Grosholz 2015: 1–7). The goal of finding a 'unified theory' may not always be the right 'regulative ideal' (to use Kant's term) for science. It may tempt scientists to look for the wrong thing, and to misconstrue what they themselves are doing. Instead, scientists should look for the best way to deploy disparate but well-correlated and well-mediated ways of representing, as they refer to and analyze a world that they both encounter and construct. The search for a unified theory of physics in the 21st century, like the search for a resolution between Newton and Leibniz, may be misconceived. This insight shapes the way in which I understand the episodes that occupy the next sections.

In Chap. 2 of *The Dappled World*, Nancy Cartwright argues, "First, a concept that is abstract relative to another more concrete set of descriptions never applies unless one of the more concrete descriptions also applies. These are the descriptions that can be used to 'fit out' the abstract description on any given occasion. Second, satisfying the associated concrete description that applies on any particular occasion is what satisfying the abstract description consists in on that occasion" (Cartwright 1999: 49). Thus the more concrete and the more abstract descriptions are compatible; they can both coherently be attributed to the same situation or event at the same time. Indeed, to make the abstract description effective or explanatory, it must be compatible with a concrete description: abstract descriptions can only be used to say true things if they are combined with concrete descriptions that fix their reference in any given situation. However, this doesn't mean that both descriptions can be re-inscribed in a discourse constituted by a single formal language, via the kind of translation required by Nagel's account of theory reduction, discussed in Chap. 4.

To block this assumption, Cartwright adds the caveat, "The meaning of an abstract concept depends to a large extent on its relations to other equally abstract concepts and cannot be given exclusively in terms of the more concrete concepts that fit it out from occasion to occasion" (Cartwright 1999: 40). The more abstract description adds important information that cannot be 'unpacked' from any or even many of the concrete descriptions that might supplement it from our awareness of the things successfully referred to; there is no sum of referential descriptions that relieves discourse of the task of analysis. Conversely, I would add, the more concrete descriptions have meanings of their own that are to a large extent independent of the meaning of any given abstract term they fall under; there is no complete sum of the conditions of intelligibility of a thing that will relieve discourse of the task of referring. The Nagelian ideal of theory reduction, as I argued in Chap. 4, does not capture how scientists (or mathematicians) actually do science (or mathematics); and the attempt to achieve homogeneity would erase the useful heterogeneity among registers of discourse, and their combination, that in fact *increases* the information available in any given situation.

3 Thermodynamics and Boltzmann

Time might seem to be an odd choice of scientific topic. How can we refer to, or be aware of, time? How can time be presented or present or 'there'? Is there indeed a single 'flow' of time? Formal or material unity and spatial location are often very helpful for reference, but it is hard to see how they would help us refer to time. Likewise, how can we analyze time? What makes time what it is—what are its requisites? What conditions of intelligibility or possibility might time have? In a sense, time doesn't seem to lend itself to thought. As a principle of sheer transience, as a principle of radical asymmetry, as a principle of unpredictability, and as infinite, time seems inimical to discourse. Moreover, like space, it cannot be perceived; whether its effects can be perceived is another difficult question. When things alter, or start up, or deteriorate, or repeat, we explain those changes by reference to the causal and material or psychological properties of things in time, not time itself.

Conversely, discourse seems inimical to time. When we analyze by searching for conditions of intelligibility, we typically abstract from temporality in the thing investigated. We look for logical or causal conditions that will hold universally for that kind of thing: such kinds and relations are taken to be constant and general. Similarly, the more concrete representations that help us to refer also have little temporal spread: they may be schemata or descriptions that would help us identify an 'x' any time an 'x' shows up. The conceptualization that leads to discourse pulls away from time and towards the timeless, transcendent, eternal, atemporal, divine. (Philosophers disagree about what should be opposed to time). All the same, there are many scientific and mathematical discourses that seem historically to have advanced our understanding of time, so it should be helpful to look at them in more detail.

Given the preceding discussion in this book (after all, number and figure are no more or less elusive than time), we may expect two registers of discourse in scientific investigations of time. The more 'analytic' register will look at time as a condition of intelligibility for other things, and treat it in terms of pure mathematics (arithmetic and geometry), so that its relations with other equally abstract notions can be developed and exploited. The more 'referential' register will look at time in relation to the concrete descriptions that 'fit it out' on various occasions, and which study it as a constitutive feature of objects like clocks and dynamical systems. Then we may notice that these distinct modes of representation are compatible in many contexts, but exhibit interesting tensions and incompatibilities in other contexts. Both systematic compatibility and disruptive incompatibility may contribute to the growth of scientific knowledge.

All the laws of Newtonian mechanics are time reversal invariant; if time is indifferent to physical systems, they should also be indifferent to time. A century after Newton, Pierre-Simon Laplace wrote that if all the particles in the universe somehow suddenly retained their position but reversed the direction of their velocity, the universe would run backwards, because the laws of Newtonian mechanics are time reversal invariant. Many people took that claim to mean that time could just as easily run backwards as forwards, and that—t might be substituted for t in the great equation of the universe.

What it really means, however, is that if we have assigned Cartesian coordinates to the timeline, we could say by convention that the beginning of our process is $t = -T$ and that the relevant process runs to $t = 0$, so that we go from state $x(-T)$ to state $x(0)$. (By state $x(t)$, I mean that both the position and velocity of each particle is specified.) At $t = 0$, however, we must initiate a different process (that means, in the case that Laplace envisages, a different universe), in which the particles are in the same configuration as $x(0)$ but the velocities are reversed; we can call it $x'(0)$. Then in the interval between $t = 0$ and $t = +T$, this second universe will evolve from state $x'(0)$ to state $x'(T)$, in which all the particles will be positioned like those of $x(-T)$ but with the velocities reversed. Thus clearly we cannot identify $x'(0)$ with $x(0)$ and we also cannot identify $x\ (-T)$ with $x'(T)$. Time does not run backwards, and indeed *our* universe does not run backwards: it is in another (formal) universe that a process takes place that (formally) reverses a process that took place in ours. It is this formal possibility that misleads one into thinking that Newtonian mechanics allows for the possibility of time going backwards, or for a process in the universe, or the universe itself, running backwards (see Earman 2002; Sachs 1987).

However, my argument here about the meaning of Laplace's thought experiment may not dissuade the Newtonian scientist, who would like analytic time to render superfluous any appeal to referential time. Perhaps the symmetry of the line, or the time symmetric invariance of the equations of motion, is a better guide to time than the apparent empirical asymmetry of our experience; for a Newtonian, time must be investigated as a condition of intelligibility, and thus by formal rather than empirical means. And the time reversal invariance of Newton's equations also applies to Einstein's equations of motion in both Special and General Relativity Theory, and to Schroedinger's equation and Dirac's equation in Quantum Mechanics. All the same, 19th c. Thermodynamics and certain developments in 20th c. Quantum Mechanics as well as Relativistic Quantum Field Theory pose serious challenges to this understanding of time, for they appear to offer physics certain laws and rules that are not time reversal invariant.

The modern discipline of thermodynamics, the study of heat, begins in 1824 with the publication of Sadi Carnot's *Réflexions sur la puissance motrice du feu*, considered the source for the laws of thermodynamics. The second law of thermodynamics states that the entropy of an isolated system not in equilibrium will tend to increase over time. Following the work of Rudolf Clausius and James Clerk Maxwell, Ludwig Boltzmann and J. Willard Gibbs defined entropy in a way that led to their definition being interpreted (not without controversy) as a measure of disorder. Clearly this law is not time reversal invariant, for an increase in entropy corresponds to irreversible changes in a system. But is this law analytic or referential? That is, does this law report a condition necessary for the intelligibility of physical processes, or is it a very high-level empirical generalization? And what does this law reveal about the nature of time? (See Müller 2007; Carroll 2010; Penrose 2011.)

In the mid-nineteenth century, the second law of thermodynamics was a notable exception within physics; today, it rules out decreases in the entropy of closed systems (except for the occasional statistical fluctuation), and therefore by extension in the entropy of the cosmos as a whole (depending on how the 'wholeness' of the cosmos is conceived). However, the reduction of thermodynamics to statistical mechanics carried out by Boltzmann, Maxwell and Gibbs at the end of the 19th century explained away this apparent asymmetry in macroscopic processes (T-violation) by reformulating them in terms of time-reversal invariant microscopic processes with special initial conditions: heat phenomena can be reduced to and redescribed as the collisions of large numbers of particles, governed by the laws of Newtonian mechanics. At the macroscopic level, an ice cube in a glass of water at room temperature will always melt; however, another possible solution to the Newtonian equations of motion that describe the situation of the particles in the glass is that an ice cube spontaneously forms out of the water. This seems to violate the second law of thermodynamics, but Boltzmann explains the apparent paradox by noting that while this 'solution' is logically possible it is statistically extremely unlikely (see Horwich 1987; Cercignani 2006; Zeh 2007). The second law of thermodynamics is a statistical regularity, not a universal and necessary law like the laws of Newtonian mechanics. We are misled by the fact that we apprehend events at the macroscopic level via the approximations provided by our sense organs.

This argument seems plausible when the example is a glass of water, but less so when the example is the cosmos. Scientists want to explain events as well as to describe them. If all the laws of physics were truly time reversal invariant, the macrostates of the past should be no more and no less probable than those of the future; but then the thermodynamical evolution of the cosmos seems inexplicable. Since most scientists agree that entropy has been increasing as the cosmos evolves, science must furnish an explanation of this fact. Roger Penrose hoped to offer an explanation by postulating an initial condition for the cosmos of very low entropy, cast in terms of a new concept of gravitational entropy (Penrose 1979). However, even if that explanation does not develop as planned, one would apparently have to conclude that time reversal invariance cannot characterize all the fundamental laws of physics; other laws that explain the universal asymmetry must be added, and the second law of thermodynamics seems a likely candidate. Boltzmann's reduction demotes the second law to merely referential status; but this strategy undermines the achievement of one of science's most important analytic tasks, explanation.

4 General Relativity and Quantum Mechanics

The seeds of Einstein's revolution lay in James Clerk Maxwell's set of partial differential equations that laid the foundation of classical electrodynamics and optics. Einstein postulated the invariance of the velocity of light c in a vacuum as a result of studying Maxwell's results, and then thought through the consequences of this insight in relation to Galilean relativity (the claim that inertial motion, straight line motion at a constant speed, is physically equivalent to rest). In his Theory of Special Relativity (1905), the first of his revolutionary theories, he argued that Newton's notions of absolute time as well as of absolute simultaneity were untenable. While time in the Theory of Special Relativity was no longer absolute, however, the space-time framework was still rigid, a fixed 'stage' within which physical events simply take place. Based on the failure of scientists to detect any changes in the relative velocity of light (as they attempted to detect motion relative to the aether), Einstein posited that the speed of light in a vacuum c will be the same for all inertial frames, and thus the relative speed between moving bodies could never be accelerated beyond the speed of light. He realized that if we pursue consistently the implications of Galilean invariance (all uniform motion is relative and there are no privileged inertial reference frames) together with the invariance of the speed of light, consequences unforeseen by Newtonian mechanics—and of course barred by Newton's insistence on absolute space and time—must follow. The duration of the time interval between two events, and what counts as a set of simultaneous events, for example, will not be the same for all observers, so that talk of space and time must become talk of 4-dimensional spacetime; and energy and mass will be equivalent, that is, $E = mc^2$ (Ashtekar 2015).

Einstein's second revolution, the Theory of General Relativity (1915), resulted from his attempt to unify special relativity with a theory of gravity. Einstein was motivated in this work by two seemingly simple observations. Ashtekar notes: "First, as Galileo demonstrated through his famous experiments at the leaning tower of Pisa, the effect of gravity is universal: all bodies fall the same way if the only force on them is gravitational. Second, gravity is always attractive. This is in striking contrast with, say, the electric force where unlike charges attract while like charges repel. As a result, while one can easily create regions in which the electric field vanishes, one can not build gravity shields. Thus, gravity is omnipresent and non-discriminating; it is everywhere and acts on everything the same way. These two facts make gravity unlike any other fundamental force, and suggest that gravity is a manifestation of something deeper and universal" (Ashtekar 2015). Since space-time is also omnipresent, Einstein came to see gravity and space-time as expressions of the same cosmic structure: space-time is curved by the presence of matter and it is dynamic; the effects of gravity are explained by the curvature of space-time. So time in general relativity becomes 'elastic': whereas in special relativity time intervals depend on the motion of the observer, in general relativity they can also depend on the location of the observer, because of variable curvature. The evolving geometry of space-time opens up the possibility of singularities. In particular, if one assumes that there is no preferred place or preferred direction in the universe, the hypothesis of the Big Bang follows from Einstein's equations, as Georges Lemaître recognized even before Edwin Hubble collected his revolutionary astronomical data. On this hypothesis, the universe would have begun at a finite time in the past, and it would then be meaningless to ask what happened before: there would be no earlier time, for time would have had a beginning (Ashtekar 2015).

The latest revolution involves the ongoing attempt to unify general relativity with quantum mechanics, since the earliest stages of a cosmological model of the Big Bang entail a very high energy density, whose features can only be understood in terms of the relativistic quantum properties of energy in the Planck regime: quantum energy is described by quantum fields (see Weinberg 1977/1993). Moreover, the predictions of general relativity for those early stages are so far physically meaningless or unreliable. Current efforts to unify general relativity and quantum mechanics, however, sometimes suggest that space-time might not be a smooth continuum, and sometimes suggest that time might not be a single flow; and this of course complicates our conception of time. The most successful account of time in this context, according to Jeremy Butterfield, is a reprise of relational time: one chooses a suitable, monotonically increasing dynamical variable for the cosmos (the reciprocal of total matter density, for example, or a suitable curvature scalar) and uses it as a clock with respect to which all other observables change. The problem is then that in quantum gravity, the graininess of space-time blocks the foliation of space-time into space-like slices and thus the consistent reconciliation of time as defined in terms of distinct choices of clock-variables (as occurs in general relativity). One hopes for a new kind of congruence among choices to emerge, but what its nature will be and whether it is really possible remains to be seen (see Butterfield 2015).

Other controversial aspects also arise. Quantum theory, the collective creation of perhaps a dozen physicists in the early 20th century, seems to entail (if one takes state reduction seriously) an asymmetry in time. (Some physicists and philosophers, for example David Deutsch, Leonard Susskind and Simon Saunders, contest this claim, but it is widely held: Roger Penrose, Giancarlo Ghirardi and others champion it.) Although Schrödinger's equation, which describes how the quantum state of a physical system changes over time, is time reversal invariant, a standard though contested tenet of quantum mechanics is that the measurement of a system irretrievably destroys information, so that we cannot reconstruct the system's initial state once we have measured it. The measurement required for scientific investigation is itself an irreversible process, and this 'state reduction' is as fundamental to Copenhagen quantum mechanics as the Schrödinger equation. (Heisenberg, Pauli, Dirac and von Neuman held this view, but it was opposed early on by Schrödinger and Einstein.) (Ney and Albert 2013). A striking consequence of the two 20th century revolutions in physics is thus that general relativity (whose best empirical verification occurs on a cosmic scale), Newtonian mechanics (which works best on and around the solar system), and quantum mechanics (which most successfully explains very small-scale phenomena) seem logically incompatible in many respects, not least a propos their accounts of space and time.

Just as Boltzmann wanted to demote the asymmetry of entropy, so a research program based on the notion of quantum decoherence (a program active since the 1980s) dismisses the asymmetry of state reduction as an illusion that stems from treating any system as isolated. According to this account, since no system is truly isolated, the information lost in the process of measurement is not really destroyed, but leached away into the ambient environment, where it is dispersed. So the apparent in-principle irreversibility of the process is replaced by the statistically unlikely possibility of retrieving the dispersed information from the environment in which it has become entangled. This approach however generates the following quandary: if when the measurement is made the information isn't really lost, then according to quantum mechanics in a sense a definite outcome of the measurement has not taken place. Thus Roger Penrose and some of his colleagues are pursing a program opposed to the quantum decoherence program, according to which state reduction and its attendent time asymmetry must be real (see Penrose 2004: Chaps. 22, 29 and 30; Schlosshauer 2008; Jaeger 2009). Other physicists, including Lee Smolin, as noted above, also defend the reality of the arrow of time.

5 Coda

The attempt to bring general relativity theory and quantum theory into closer relation in 20th (now 21st) century cosmology has altered our understanding of the physical world. At first, general relativity was supposed to refer to cosmological and astrophysical phenomena, and quantum mechanics to the study of atomic and subatomic particles. However, in the states of the big bang and the big crunch or bounce, where the cosmos is hypothesized to be very dense, the world of general relativity and quantum mechanics must meet, and scientists wish that at that meeting point their descriptions and predictions would be at least compatible.

The electromagnetic field was successfully quantized by representing the dynamical field, a tensor field $F_{\mu\nu}$ on Minkowski space-time, as a collection of operator-valued harmonic oscillators (see Schweber 1994). One might say that Minkowski space-time and the isometries of the Minkowski metric play an analytic role here, and constructed physical quantities such as fluxes of energy, momentum and angular momentum (carried by electromagnetic waves) play a role that is more concrete and referential. However, it has proved much harder to quantize gravity and general relativity. In general relativity, there is no background geometry; the space-time metric itself is the fundamental dynamical variable. As Abhay Ashtekar writes, "On the one hand, it is analogous to the Minkowski metric in Maxwell's theory; it determines space-time geometry, provides light cones, defines causality, and dictates the propagation of all physical fields (including itself). On the other hand it is the analogue of the Newtonian gravitational potential and therefore the basic dynamical entity of the theory, similar in this respect to the $F_{\mu\nu}$ of the Maxwell theory. This dual role of the metric is in effect a precise statement of the equivalence principle that is at the heart of general relativity. It is this feature that is largely responsible for the powerful conceptual economy of general relativity, its elegance and aesthetic beauty, its strangeness in proportion. However this feature also brings with it a host of problems… We see already in the classical theory [of General Relativity] several manifestations of these difficulties. It is because there is no background geometry, for example, that it is so difficult to analyze singularities of the theory and to define the energy and momentum carried by gravitational waves. Since there is no a priori space-time, to introduce notions as basic as causality, time, and evolution, one must first solve the dynamical equations and *construct* a space-time" (Ashtekar 2012: 2).

How do scientists address this problem? There were two approaches, according to Ashtekar's historical summary, the canonical and the covariant, a late 20th c. contest that reproduced, again, the tension between a dominance of the abstract formulation on the one hand, and the concrete formulation on the other. The canonical approach aimed to retain Einstein's fusion of gravity and geometry. It used the Hamiltonian formulation of general relativity, and designated the three-metric on a space-like slice as the basic canonical variable. Despite its commitment to the analytical and geometrical, it was led to internally differentiate: the set of ten Einstein's equations were decomposed into two sets, four (more analytic) constraints on the metric and its conjugate momentum, and six (more referential) evolution equations. This research program became stagnant, however, since the use of this framework in the study of particle physics produced very difficult, perhaps unsolvable, equations. In addition, concepts and methods that had proved successful in quantum electrodynamics (scattering matrices, Feynman diagrams) did not emerge here, so that practitioners of particle physics could find little common ground.

Another more referential approach, the covariant approach, thus came to seem more promising; despite its commitment to the concrete—the dynamical object—it was also forced to assign distinct roles to distinct parts of discourse, but in a different way. It split the space-time metric itself into two parts, the first more analytic and the second more referential. Thus, $g_{\mu\nu} = \eta_{\mu\nu} + \sqrt{G}h_{\mu\nu}$, where $\eta_{\mu\nu}$ is a background, kinematical metric (often chosen to be flat), G is Newton's constant, and $h_{\mu\nu}$ is the deviation of the physical metric from the chosen background, the dynamical field; Ashtekar observes, the two roles of the metric tensor are now split. As this framework is applied to the study of particle physics, in the transition to the quantum theory, only $h_{\mu\nu}$ is quantized; quanta of this field propagate on the classical background space-time with metric $\eta_{\mu\nu}$. If the background is chosen to be flat, the quanta turn out to be gravitons, the machinery of perturbation theory and the Feynman rules apply, and there is a way to compute amplitudes for scattering processes. This approach, however, is only applicable if gravitation is weak (Ashtekar 2012).

One reflection that this discussion brings to mind is that time seems as resistant to the idioms of science and mathematics as it is to those of philosophy; yet we must go on trying to talk about it. All things arise and end in time, even the cosmos itself. Time still seems (to me) to be both transcendent and fatal, a principle of transience that resists every kind of discourse. This is odd, since we usually think of transcendence as our refuge from fatality. What do these disputes tell us about time? A Leibnizian physicist might exclaim, "So what if all fundamental physical laws are time reversal invariant! That doesn't prove that nature with its 'real time' *is not* asymmetric!" By the same line of argument, I suppose that if some of the fundamental physical laws turn out *not* to be time reversal invariant, that development also wouldn't prove that nature with its 'real time' *is* asymmetric. A Newtonian physicist might respond, "So what if the whole measureable history of the cosmos exhibits an important asymmetry (the increase in entropy)! That doesn't prove that time itself *is* asymmetric!" By the same line of argument, I suppose that if we discover that a new empirical method makes that apparent asymmetry disappear, that development also wouldn't prove that time itself *is not* asymmetric. Adherence to the analytic Newtonian ideal on the one hand, or adherence to the referential Leibnizian ideal on the other, seems to lead to skepticism about the nature of time. My suggestion is that our best hope of investigating time is to seek to combine disparate theories, some better designed for analysis, some for reference, without forcing them into an artificial unity, in order to construct a novel and more effective integration. The result might look, surprisingly, like philosophical dialectic; and yet it might still be good science.

References

Ashtekar, A. (2012). Introduction to loop quantum gravity. In *arXiv*: 1201.4598v1 [gr-qc] 22 Jan. 2012: 1–30.

Ashtekar, A. (2015). Time in fundamental physics. In E. Grosholz (Ed.), *Time and Cosmology, Special Issue of Studies in History and Philosophy of Modern Physics, 52,* 69–74.

Butterfield, J. (2015). Assessing the montevideo interpretation of quantum mechanics. In E. Grosholz (Ed.), *Time and Cosmology, Special Issue of Studies in History and Philosophy of Modern Physics, 52*, 75–85.

Carroll, S. (2010). *From eternity to here: The quest for the ultimate theory of time*. Cambridge, MA: MIT Press.

Cartwright, N. (1999). *The dappled world.* Cambridge: Cambridge University Press.

Cercignani, C. (2006). *Ludwig Bolzmann: The man who trusted atoms*. Oxford: Oxford University Press.

De Gandt, F. (1995). *Force and geometry in Newton's Principia*. Princeton: Princeton University Press.

Earman, J. (2002). What time reversal invariance is and why it matters. *International Studies in Philosophy of Science*, *16*(3), 245–264.

Grosholz, E. (Ed.) (2015). *Time and Cosmology, Special Issue of Studies in History and Philosophy of Modern Physics, 52.*

Horwich, P. (1987). *Asymmetries in time: Problems in the philosophy of science*. Cambridge, MA: MIT Press.

Jaeger, G. (2009). *Entanglement, information, and the interpretation of quantum mechanics*. New York: Springer.

Müller, I. (2007). *A history of thermodynamics*. New York: Springer.

Newton, I. (1934). In A. Motte (Tr.) & F. Cajori (Ed.), *Mathematical principles of natural philosophy and his system of the world*. Berkeley: University of California Press.

Ney, A., & Albert, D. (Eds.). (2013). *The wave function*. Oxford: Oxford University Press.

Penrose, R. (1979). Singularities and time-asymmetry. In W. Hawking & W. Israel (Eds.), *General relativity: An Einstein centenary* (pp. 581–638). Cambridge: Cambridge University Press.

Penrose, R. (2004). *The road to reality: A complete guide to the laws of the universe*. New York: Knopf.

Penrose, R. (2011). *Cycles of time: An extraordinary new view of the universe*. New York: Vintage.

Sachs, R. G. (1987). *The physics of time reversal*. Chicago: University of Chicago Press.

Schlosshauer, M. (2008). *Decoherence and the quantum-to-classical transition*. New York: Springer.

Schweber, S. S. (1994). *QED and the men who made it: Dyson, Feynman, Schwinger and Tomonaga*. Princeton: Princeton University Press.

Smolin, L. (2013). *Time reborn*. New York: Houghton Mifflin Harcourt.

Smolin, L. (2015). Temporal naturalism. In E. Grosholz (Ed.), *Time and Cosmology, Special Issue of Studies in History and Philosophy of Modern Physics, 52*, 86–102.

Weinberg, S. (1977/1993). *The first three minutes: A modern view of the origin of the universe*. New York: Basic Books.

Zeh, H. D. (2007). *The physical basis of the direction of time*. New York: Springer.

Chapter 8
Analysis and Reference in the Study of Astronomical Systems

Scientific language used in the study to elaborate and systematize abstract thought is often very different from language used by scientists working in the laboratory, field and observatory. The chemist must bring the equations of quantum mechanics into relation with the records of experimental processes, as well as diagrams and computer simulations of a given molecule in its experimental setting. The botanist must integrate the high-level principles of neo-Darwinian theory with her field records, genetic information and statistical representations of the fate of her plants. Texts that announce important ideas, bringing two or more spheres of activity into intelligible relation, are therefore typically heterogeneous and multivalent. But philosophers who begin with logic, as we have noted, seem to assume either that the reports of empirical observation can be re-written in the same formalized idiom as the first principles of the scientific theory, because they are ultimately deducible within the theory, or that they can be re-written as a structure that corresponds isomorphically to the theory. Assumptions about the logical homogeneity of scientific discourse have not been directly challenged by philosophers known for other kinds of challenges to logical positivism, like Kuhn, Popper, Van Fraassen and Kitcher.

Philosophers of science need to ask new questions that bring the work of combination itself into focus. Is there a useful taxonomy of strategies of combination across mathematics and the sciences? How does it contribute to the growth of knowledge? In this chapter I will focus on these two questions, and illustrate my reflections on them by referring first to Newton's *Philosophiae Naturalis Principia Mathematica*, the *Principia*, and then to the ensuing debates about the structure of the solar system and, later, of galaxies and the cosmos. In his essay "Mathematics, Representation and Molecular Structure", Robin Hendry notes that Nancy Cartwright (and Margaret Morrison) distinguish strongly between two kinds of models (Hendry 2001). On the one hand, philosophers like Bas van Fraassen, in his earlier book *Laws and Symmetry*, pay most attention to theoretical models, which as in model theory are structures that satisfy a set of sentences in a formal language:

E.R. Grosholz, *Starry Reckoning: Reference and Analysis in Mathematics and Cosmology*, Studies in Applied Philosophy, Epistemology and Rational Ethics 30, DOI 10.1007/978-3-319-46690-3_8

such structures are themselves organized as a language, so that the sentences of the formal language are true when interpreted in terms of the object-language. On the other hand, philosophers like Cartwright and Morrison remind us of the importance of representational models, where the relevant relation is not satisfaction (as between a meta-language and an object-language), but representation (as between a discursive entity and a thing that exists independent of discourse), like the iconic images that represent molecules (this one is benzene, C_6H_6, in Fig. 1).

Different models, or modes of discourse bring out different aspects of the 'target system'. As I have argued, those that help us to elaborate theory and the abstract, highly theoretical networks that lead to scientific explanation, typically differ from those that help us to denote, to single out the intended 'target system'. The relation between metatheory and object theory is isomorphism; but isomorphism leaves us adrift in a plurality of possible structures; most of the time, scientists cannot allow themselves to drift in that way. Hendry writes, "… we note that equations are offered not in isolation, but in conjunction with text or speech. This linguistic context is what determines their denotation and serves to make representation a determinate, non-stipulative relation that may admit (degrees of) non-trivial success and failure. Natural languages like English, French or German equip their speakers with abilities to refer to their surroundings, and we can understand how equations can represent if they borrow reference from this linguistic context" (Hendry 2001). In sum, theoretical models by themselves are too general; they cannot help us refer properly to the things and systems we are trying to investigate. And referential models by themselves are too limited; they cannot offer the explanatory depth that theory provides.

My intention in this chapter is to show how models of the solar system, our galaxy and closest galaxy-neighbor Andromeda, and our cosmos have historically proved to be composites: in order to be effective, they must combine analytic and referential modelling in an uneasy but fruitful unity. We need a thoughtful account of the variety of strategies that scientists use to construct such composite models. The relative stability of successful models makes scientific theorizing possible; and

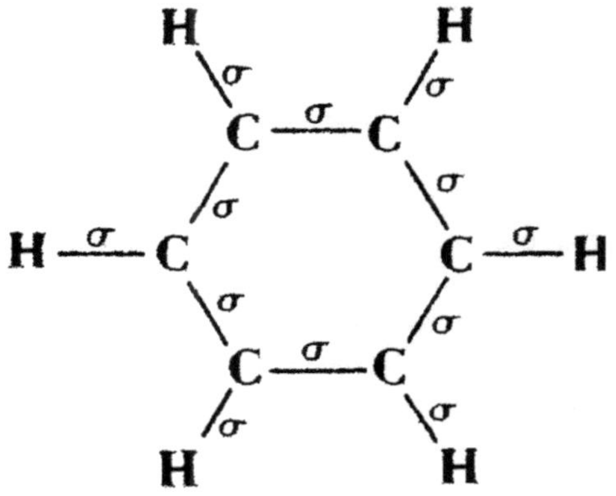

Fig. 1 Benzene Molecule

the residual instability (which no logician can erase) leaves open the possibility of further analysis and more accurate and precise reference. Models are revisable not only because they are 'approximations' that leave out information, but also because they must combine both reference and analysis. To inquire into the conditions of intelligibility of formal or natural things, we may decompose them in various ways, asking what components they have and how those components are related, or asking what attributes go into their complex concepts. We can also ask what laws they satisfy. But in order to refer properly to a thing or system, we have to grasp it first as what it is, a canonical unity or a whole relatively stable in time and space, its oneness governing the complex structure that helps to characterize it.

Things and systems—both natural and formal—have symmetries and (since periodicity is symmetry in time) so do natural processes! Carbon molecules as they throb, snowflakes as they form, and solar systems as they rotate exhibit symmetries and periodicities that are key to understanding what they are and how they work. Thus the shape (in space and time) of a system or thing is not, as Aristotle once claimed, a merely accidental feature. On the contrary, symmetry and periodicity are a kind of generalization of identity; they are the hallmark of stable existence. Symbolic modes of representation seem to be most useful for abstract analysis, and iconic modes of representation for reference: a representation of shape is often an important vehicle for referring. This is an over-simplification, however; tabulated data and data displayed to exhibit (for example) linear correlations have both symbolic and iconic dimensions, and most icons come equipped with indices that relate them to symbolic notation. Thus we should expect models to be both symbolic and iconic. And then it is rewarding to ask, how do those modes of representation interact, on the page and in thought?

1 A Taxonomy of Strategies for Integrating Reference and Analysis: Newton's *Principia*

If the tasks of analysis and reference are often disparate, we may expect that, for example, the records kept by astronomers (even when their work is informed by theory) will differ from the theorizing of physicists and natural philosophers (even when they are concerned primarily with celestial systems). The modes of representation and the idioms of mathematical and scientific expression will differ; and the explication and organization required of natural language will differ from one task to the other. What we can then expect are composite texts in both kinds of endeavor; and the nature of this composition can suggest a preliminary taxonomy of 'strategies of integration'. I will illustrate each one in terms of Newton's *Principia*.

The simplest such strategy is juxtaposition. The overall structure of the *Principia* is a striking example of juxtaposition. Book I presents Newton's mathematical method for analyzing physical phenomena, and Book III is the application of those methods to the solar system (Book II is in part a refutation of Cartesian vortex mechanics, which I won't discuss in this chapter). What we find in Book I are geometrical diagrams adjusted to accommodate Newton's 'method of first and last ratios' which makes possible the analysis of motion in terms of position, time, velocity, force and mass. The paradigm of motion is the orbit of a body around a center of force, introduced in Proposition I, and the culmination is the proof of the inverse square law for gravity in Proposition XI; clearly, the canonical object for Newton in Book I is the solar system. Despite the strict geometrical abstraction of Book I, in the Scholium to Proposition IV Newton mentions Wren, Hooke, Halley, and Huygens, adding "in what follows, I intend to treat more at large of those things which relate to centripetal force decreasing as the squares of the distances from the centres" (Newton 1999: 452). In Book III, by contrast, Newton includes many tables copied from the records of various astronomers, and he insists on the congruence of records kept by different astronomers in different countries who are tracking the same celestial object. Book III also includes two stylized pictures of the transit of a comet through the heavens (Newton 1999: 906 and 919). Its geometrical diagrams differ strikingly from those in Book I, because they show techniques for adjusting the predictions of theory to the complexities of real astronomical data (for example, the moon is the earth's satellite, but the influence of the sun is considerable and must be factored in).

A variant of juxtaposition is the kind of two-dimensional correlation that produces tables, systems of equations, matrices, and graphs. In a manner that is rather subtle, since these arrays are present to the eye on the page all at once, a two-dimensional array presents two different kinds of information at once, locating the terms in two different kinds of discourse. In Galileo's account of projectile motion in the Fourth Day of the *Discorsi*, as we saw, he juxtaposes horizontal inertial motion (which is a creature of theory) with vertical free fall (which is the mathematically elaborated observational description of how bodies fall near the earth); their compounding produces the parabola of the trajectory of a projectile. And Newton borrows this strategy in Book I, Proposition I of the *Principia*: at each moment of its motion, a body revolving around a center of force endeavors to continue in inertial motion, but is deflected into free fall by an impulse of the centripetal force. The further juxtaposition of these moments side by side in the diagram, governed by Euclidean rules about the equality of the area of triangles and rendered infinitesimalistic and so dynamic by the method of first and last ratios, produces a demonstration of Kepler's law of areas (Newton 1999: 444–446).

A more nuanced form of juxtaposition is best explained by beginning with the venerable notation of proportions. Euclid's treatment of ratios and proportions in Book V of the *Elements* provides a useful schema. For Euclid, ratios hold only between things 'of the same kind'. Proportions that yoke two ratios, $a : b :: c : d$, thus allow mathematical discourse to bring different kinds of things into relation without denying their disparity: though numbers are different from line segments, the ratio between two numbers may be considered analogous to the ratio between two line segments. The conclusion of the proof of the inverse square law, Book I, Proposition XI, is just such a proportion: $\mathrm{L} \times \mathrm{QR} : \mathrm{QT}^2 :: 2\mathrm{PC} : \mathrm{G}v$. The magnitudes on the left-hand side have physical import, resulting from the expression of the dynamical features of the idealized physical system; the magnitudes on the right-hand side are finite geometrical line lengths with no physical import, stemming from the mathematical properties of the figures involved (Newton 1999: 262–263; see also Grosholz 1987: 209–220).

A modern elaboration of this strategy is the notion of isomorphism: two domains are acknowledged to be different in kind, but their structural similarities can be captured by an isomorphism, a one-to-one mapping that preserves structure. The schematic picture of Halley's comet of 1680 presented in the 'example' following Proposition XLI, Book III (based upon the tabulated data given on the preceding pages), and the geometrical diagrams in Proposition XLI that exhibit its trajectory as a parabola, share a common structure, which constitutes for Newton significant evidence of the truth and fruitfulness of his theory (Newton 1999: 906 and 902).

Some kinds of integrative strategies, however, we may call unification in the logicians' sense, for they insist on a stronger rapprochement between the two disparate epistemic tasks and the resultant disparate discourses. Euclid also studies continuous proportions, $a : b :: b : c$. In such proportions, the sharing of the term b between the two ratios guarantees that a, b and c are all the same kind of thing. For Newton, spatial position, temporal duration, velocity, mass and force are very different kinds of magnitudes—the latter pertain to the dynamical object and the former to the metaphysical framework furnished prior to objects by God. Yet for certain purposes of calculation he treats them all as numbers, and then the proportions that link them can be treated by special methods formerly reserved (in the Middle Ages) for continuous proportions. Specifically, the proportion is arithmetized, ratios become fractions (a new kind of number), the compounding of ratios can be treated as the multiplication of fractions, and analogy becomes an equality: $a/b = c/d$ (Sylla 1984). That is, the unification of Newtonian mechanics depends not only on the discovery of first principles and the constitution of an axiomatized theory; it also depends on the re-formulation of things as if they were generalized objects, or rather on the substitution of generalized objects for things. Thus magnitudes are treated as if they were numbers (with appropriate units), the large objects of the solar system are treated as if they were point masses in one sense and a sequence of spatio-temporal points in another, and trajectories become, in the Leibnizian development of Newtonian mechanics, solutions to differential equations.

Re-writing the things of a domain to reduce the disparity between the task of analysis and the task of referring has a number of aims. A scientist like Newton who is looking for a unified physics is motivated to minimize the differences among things not only so that they can be treated by one unified theory, but also so they can be combined in novel ways, subjected to calculation, and made amenable to general methods. We recall how Karine Chemla stresses the importance of 'modes of application' in mathematics, which allow processes of generalization, in relation to Desargues' projective geometry and to classical Chinese mathematics (see Chemla and Shuchun 2004). One such example is Descartes' re-writing of algebraic curves as equations, so that his rule for finding tangents becomes immediate and straightforward. The effort to extend Descartes' general methods to 'mechanically generated' transcendental curves by Leibniz and the Bernoullis led to the elaboration of notation for equations with variable exponents, and special symbols like *sin*, *cos*, *e* and *log*, notation for infinite series, and notation for differential equations.

Ernst Cassirer, as we saw in Chap. 1, explores a strong version of this strategy of unification in *Substanzbegriff und Funktionsbegriff*, focusing on how Descartes was able to exhibit the systematic connections among conic sections via his re-writing of curves into equations. Arguing against the Aristotelian account of abstraction and a naïve view of induction, Cassirer claims that the great conceptual innovation of early modern science was to seek mathematical concepts that do not cancel or forget the determinations of the special cases, but fully retain them, so that they can be deduced from the universal formula. If a general concept had been arrived at by Aristotelian abstraction, the special cases could not be recovered from it, because the particularities would have been forgotten. By contrast, the mathematical or scientific concept seeks to explain and clarify the whole content of the particulars by exhibiting their deeper systematic connections, revealed in the law of the series. Here the more universal concept is more, not less, rich in content; it is not a vague image or a schematic presentation, but a principle of serial order. Thus in modern mathematics, things and problems are not isolated, but shown to exist in thoroughgoing interconnection (Cassirer 1923/1953: Chap. 3).

Moreover, if the general concept had been arrived at by naïve induction, it would have emerged from disconnected particularities. The rule of the series, Cassirer argues, is not attained through bare summation; rather, its elements have already been presupposed as organized in an ordered manifold. What Newton was able to do in formulating his three laws and the law of Universal Gravitation, was to add something to the theoretical situation that had not been there before, his unprecedented treatment of force and mass. In Book I of the *Principia*, a series of things with attributes (specifically the bodies of the solar system) is transformed into a systematic totality of variable terms or parameters; things are transformed into the solutions of mathematically articulated problems. As Cassirer puts it, the world of sensible presentations is not so much reproduced as supplanted by an order of another kind. Then whenever we unify the 'first order' objects of our thought into a single system, we create new 'second order' objects whose total content is expressed in the functional relations holding among them; they can be deduced from the axioms because disparity has disappeared (Cassirer 1923/1953: Chap. 5). However, we should remember that this transformation is carried out in Book I of the *Principia*; Cassirer neglects the importance of the work carried out in Book III, which is also scientific and brings the task of analysis into more robust relation with the task of reference.

If we return to Euclid's continuous proportion, $a : b :: b : c$, another strategy of unification appears, which could be called superposition. Initially it looks like a fallacy, the fallacy of equivocation, but it has serious uses. What if a and c really are different kinds of things, whose disparity cannot be dismissed by theory or practice? Then we may use the term b to mean different (but related) things in each ratio. This strategy exploits an ambiguity in the term b; however, as I have argued elsewhere, if the ambiguity is highly structured and controlled, it may be productive for mathematics and science. It sometimes allows the scientist to find a way between the theoretical looseness of juxtaposition and the loss of reality that plagues unification. In the next section, after tracking Newton's uses of juxtaposition and unification in the *Principia*, I will end by showing where and why he uses the strategy of superposition.

2 Further Consideration of Newton's Use of These Strategies in the *Principia*

In the late sixteenth century and throughout the 17th century, the problem of reference in astronomy is just as compelling as the problem of analysis. The main object of study in that period is the solar system; as we stand on the earth, the sun and the moon are large, brilliant objects in the sky, and the planets are salient and distinctive in their movements. To refer, in one sense, all we have to do is point. But the very act of pointing out an item in the solar system is a tracking: celestial objects move, so the question of how to characterize the movement must also arise. Tracking the objects of the solar system required, in the 16th century, a compass and a sextant or quadrant; Tycho Brahe used these instruments in an unusually consistent and careful fashion, calibrating his instruments regularly and measuring their positions at small temporal intervals all along a given orbit with unprecedented accuracy. Brahe's tables of planetary motion, the *Rudolphine Tables*, meant to supplant the 13th c. *Alphonsine Tables*, were published by his collaborator Kepler after his death in 1601. His famous ellipse from the *Astronomia Nova* is given below (Fig. 2).

The exposition of the *Rudolphine Tables* shows that problems of reference are solved in relation to problems of analysis; recall that my claim is not that these problems are disjunct, but that they are logically disparate and in need of further integration. Given the way that Kepler sets up the table, he is clearly using a heliocentric system with elliptical planetary orbits. This fact is noteworthy because Tycho remained opposed to the heliocentric hypothesis till the end of his life, and he died before Kepler worked out his laws of motion: the claim that the orbit of Mars is elliptical is first published in his *Astronomia Nova* (1609). Thus the tables embody and display two theoretical challenges to Aristotelian/Ptolemaic astronomy which Tycho himself never made. Second, it was the very accuracy of Tycho's data that persuaded Kepler finally to abandon his devotion to the circle, and to search for other simple mathematical forms, at last settling on the ellipse. Unprecedented accuracy (achieved largely by scaling the instruments up in size), and frequent tracking motivated a change in conceptualization. Of course, it was the highly theoretical mathematics of Euclid and Apollonius that offered a repertoire of forms to Kepler, from which he chose the ellipse (see Voelkel 2001).

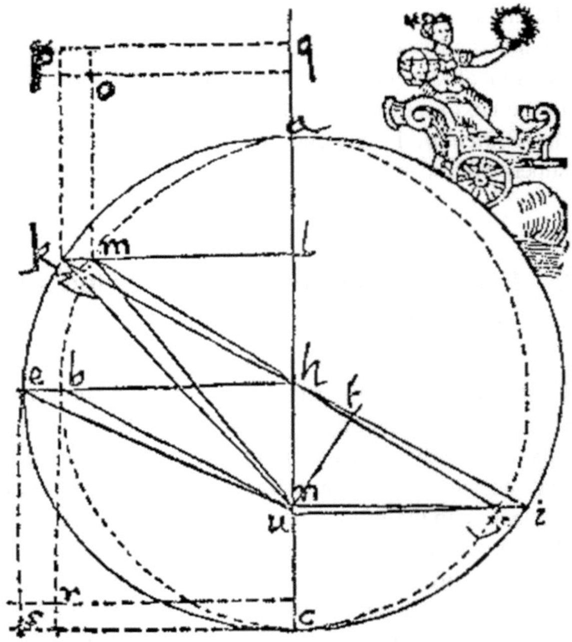

Fig. 2 Frontispiece, *Astronomia Nova*

As is well known, Galileo pounced upon the refracting telescope almost as soon as it was invented, made improvements to it, and turned it on the heavens. (Kepler was an enthusiastic supporter of Galileo's *Sidereus Nuncius* (1610), and he himself used the telescope to look at the moons of Jupiter and the surface of the moon.) Sixty years later, Newton built the first reflecting telescope, using a concave primary mirror and a flat diagonal secondary mirror; this invention impressed both Barrow and Huygens, and led to Newton's induction into the Royal Society. From then on, improvements in our ability to refer to the objects of astronomy have depended on improvements in the material composition, size, and placement of telescopes. Galaxies and galaxy clusters, if they are not simply invisible, are at first mere smudges on the night sky (a few are visible as smudges to the naked eye). Either they are not recorded, or they are noted as 'nebulae', clouds whose structure is just barely visible in the 19th century and whose composition remains mysterious until well into the 20th century. Like clouds, they seem to have no determinate shape; the discernment of galactic shape plays an important role in the development of 20th c. astronomy and cosmology (Wilson 2014: Chaps. 1 and 2).

In Book I of Newton's *Principia*, Kepler's Second Law (that planets sweep out equal sectors in equal times: they accelerate as they get closer to the sun and decelerate as they get farther away) is proved in Newtonian fashion in Proposition I, and his ellipse is the centerpiece of the diagram that accompanies the proof of the inverse square law, Proposition XI. That ellipse therefore appears as a palimpsest. It is at the same time a Euclidean-Apollonian mathematical object, with one set of internal articulations useful for discovering its mathematical properties; a tracking device for Kepler as he finishes compiling the Tables with Tycho's compass and sextant or quadrant, and therefore just an outline, since a trajectory is just a line across the sky; and finally as well Newton's construction, with a superimposed set of articulations for displaying physical and indeed dynamical properties. The ellipse is a discursive locus where the demands of reference and the demands of theory are, in Proposition XI, happily but formally reconciled (Newton 1999: 462–463). All the same, the multiple roles the ellipse is forced to play there in a sense destabilize the geometry and will ultimately lead to its re-expression in the Leibnizian form of a differential equation. In particular, if we suppose, given our observational data, that the orbiting body is subject to the effects of more than one center of force (or that the force—reaction—it exerts on neighboring bodies is non-negligible), the mathematics required is so complex that a new notation becomes imperative.

Newton's project is to explain Kepler's claim that planetary trajectories are elliptical by uncovering its conditions of intelligibility. At the opening of Book I, Newton shows that any body revolving around an immovable center of force must obey Kepler's Second Law; that result can be proved without making any assumptions about the center of force, as it is in Proposition I (Newton 1999: 444–446). In Proposition VI, Newton shows that the action of a center of force will determine the shape of the trajectory of the revolving body; thus, the shape of a trajectory will give information about the center of force (Newton 1999: 453–455). Geometry provides Newton with a wide range of possible trajectories (considered abstractly in Propositions VII–X), but only observation allows him to select as most important an elliptical orbit with the center of force at one focus (Newton 1999: 455–463). The procedures of accurate reference have generated tables that present the earth's orbit as elliptical. If the trajectory is elliptical, and the center of force is located at one focus of the ellipse, then the center of force must obey the inverse square law. Note that in the proof of Proposition XI, Newtonian analysis considers the abstract model of a solar system with only the sun and the earth (Newton 1999: 462–463).

In May of 1686, Edmond Halley wrote to Newton, to tell him that the members of the Royal Society viewed the *Principia* as such an important work, that they would undertake its printing 'at their own charge', though in fact he himself paid for its printing at some personal sacrifice. For the next year or so, much of Newton's correspondence with the distinguished astronomer concerned the printing of the *Principia*. Surprisingly, in June Newton wrote to Halley that he wanted to suppress Book III, the presentation of the system of the world. I. Bernard Cohen observes, "Halley proved to be a master of diplomacy and his saving of Book III may just possibly have been his most significant contribution to the *Principia*" (Cohen 1971: 134). Newton was worried about the reception of the book, because he was not satisfied with his studies of the motions of the moon or of the comets, but Halley persuaded him that these applications of his theory were of utmost importance to the scientific community as well as to the educated public. In July 1686, Newton complied, to the delight of Halley and the Astronomer Royal John Flamsteed.

The proof-reading and final publication of the *Principia* was completed, under Halley's direction, in July 1687 (Cohen 1971: 134–142).

Book III, *The System of the World*, begins with various admonitions from Newton to the reader. He first writes, "I chose to reduce the substance of this Book into the form of Propositions (in the mathematical way), which should be read by those only who had first made themselves masters of the principles established in the preceding Books", adding that acquaintance with the *Definitions*, the *Laws of Motion*, and the first three sections of Book I (up to the proof of the inverse square law) would suffice. Then he gives four *Rules of Reasoning in Philosophy*, and adds a notable section entitled *Phenomena*. Phenomenon I states that the moons of Jupiter obey Kepler's Second and Third laws; Newton cites in tabular form the observations of Borelli, Townly, and Cassini as well as Pound, noting that they used different methods and instrumentation. Phenomenon II makes the same point about the moons of Saturn, citing observations of Cassini and Huygens. Phenomena III, IV, and V taken together establish the same conclusion about the planets Mercury, Venus, Mars, Jupiter and Saturn, citing the results of Kepler and Boulliau in tabular form, and other astronomers, who all agree that the planets revolve around the sun (not the earth) and obey Kepler's Second Law, the Law of Areas (Newton 1999: 797–801).

Thomas Kuhn called the development of Newtonian mechanics a good example of 'normal science', the working out of puzzles nicely covered by the rules, like problems at the end of a textbook chapter. His claim is certainly true of Newton's account of the orbits of the moons of Jupiter and Saturn, and of the five planets just noted around the sun, which are handled with dispatch in Propositions I, II and V. This isn't surprising, since these very objects in just such systematic relations served as the canonical objects for Kepler and Newton: the theory was designed especially for them. Here, we see an integration of analysis and reference that looks like a deduction, an almost perfect case of the strategy of unification. But this is not the end of Book III, only the first 15 of its 150 pages! Newton's hesitation about publishing Book III stemmed from his inability to match that perfection even in the case of the moon, and for different reasons in the case of the comets. The issues raised by the moon pertained more to analysis, how to frame the 3-body problem; nobody disputed the presence of the moon or where it was in the sky. The issues raised by comets pertained more to reference: Was the comet of 1680 the same entity as the comet of 1682? Were comets really part of the solar system? Were comets made of the same stuff as the planets and their moons? Newton strives mightily in the rest of Book III to address these issues, but the proper treatment of the orbit of the moon had to wait for the new notation developed by Leibniz, the Bernoullis and Euler, and the identity of Halley's comet was not established until after Newton's death, by Halley (see Cook 1998).

In Proposition XXII, Newton asserts hopefully, "That all the motions of the moon, and all the inequalities of those motions, follow from the principles which we have laid down" (Newton 1999: 832). His attempt to justify this claim covers 60 pages, filled with geometrical diagrams (in which Newton is trying to get his geometry to do the job of differential equations), geographical reports about the tides from around the world, and a small treatise by Mr. Machin about the moon's nodes, inserted and quoted verbatim as two propositions and a scholium (Newton 1999: 832–888). His treatment of the comets runs about 50 pages, and includes geometrical diagrams, numerous astronomical tables and observations compiled by Flamsteed, Halley, Hooke and Pound, as well as observations recorded by the Italians Ponthio, Cellio, Cassini, and Montenari, the French astronomers Gallet, Ango and Bayer, the American Storer, and the German Zimmerman, and an imaginative ten-page excursus, worthy of Kipling, on the physical composition of comets and why they have tails (Newton 1999: 888–938).

In his treatment of comets, Newton must first persuade the reader that a comet belongs in the same category with the planets, that it is also a member of the solar system, a claim oddly located as a lemma following a discussion of the precession of the equinoxes, Lemma IV: "The comets are more remote than the moon, and are in the regions of the planets" (Newton 1999: 888). He spends about seven pages defending Lemma IV in terms that are strikingly qualitative, despite the presence of two diagrams. He cites the commonly acknowledged retrograde motion they share with the outer planets, and the fluctuations in the brightness of their heads as they (he infers) pass near to the sun, observed by Kepler in 1618, by Helwelcke in 1665, and by Flamsteed in 1680 and again in 1682. (Significantly, he does not identify the comet of 1680 and the comet of 1682.) He concludes, "Comets shine by the sun's light".

Having thus located comets in the solar system, he asserts in Proposition XL, "Comets move in conics having their foci in the center of the sun, and by radii drawn to the sun, they describe areas proportional to the times". He notes in Corollary I, that their orbits are ellipses, but in Corollary II, adds "but these orbits will be so close to parabolas that parabolas can be substituted for them without sensible errors", presumably because as one focus of an ellipse is drawn further and further away from the first, the ellipse becomes (in the infinite limit) a parabola, though Corollary II has no explanatory gloss (Newton 1999: 895). The geometrical treatment of the trajectory of a comet follows in seven lemmas and Proposition XLI, illustrated by nine diagrams and then by the 'example' of Halley's comet, documented in six tables and two pictures which are, like all pictures of trajectories, like time-lapse photographs. As noted above, the second picture ("a true representation of the orbit which this comet described") and the diagram-parabolas in Proposition XLI exhibit the isomorphism that Newton claims must hold between theory and the recorded trajectory.

In the midst of this exposition, Newton writes, "The observations of this comet from beginning to end agree no less with the motion of the comet in the orbit just described than the motions of the planets generally agree with planetary theories, and this agreement provides proof that it was one and the same comet that appeared all this time, and that its orbit has been correctly determined here" (Newton 1999: 911). And just before the second picture, he adds, "The theory that corresponds exactly to so nonuniform a motion through the greatest part of the heavens, and that observes the same laws as the theory of the planets, and that agrees exactly with exact astronomical observations cannot fail to be true" (Newton 1999: 916). In fact,

Newton was right about the comets; they are part of our solar system and they do correspond very well to his theory. However, my point is that the issue of the disparity of analysis and reference remains. It is evident in the contrast between Book I and Book III, but it also appears within Book I, where it is dealt with by strategies of unification, nuanced by superposition, and in Book III, where it is dealt with by strategies of juxtaposition, nuanced by isomorphism.

3 The 3-Body Problem: From Newton to Hamilton

In Book III of the *Principia*, Newton elaborates his theory and enriches his model, building in further complexity, to show that he can account for further tabular evidence compiled by other astronomers around Europe. He accounts for perturbations in the orbit of the moon in terms of the gravitational pull of both the earth and the sun, and goes on to account for the tides; he explains the orbits of comets as they intermittently visit the solar system; and he shows that not only the other planets but also the moons of Jupiter obey the generalized law of universal gravitation. The problems left for the next generation by Newton's Book III are therefore, in his opinion, 'puzzles of normal science' (Kuhn apparently concurs in Newton's assessment). On this account of scientific progress, the puzzles of reference are to locate and measure the movements of more and more astronomical items, and so to make sure that they accord with Newton's Three Laws of Motion and the Law of Universal Gravitation. Existing theory, expressed in the formal (highly geometrical) idioms of the *Principia*, will cover and explain the results of observation, and prove adequate to solving the puzzles of theory, which include first and foremost how to move from the 2-body problem to the 3-body problem to the n-body problem (Fig.6.6 is the figure given in Book I, Sect. III, Proposition XI of the *Principia*). Newton's Law of Universal Gravitation states that, in the case of two bodies, the force acting on each body is directly proportional to the product of the masses, inversely proportional to the square of the distance between their centers, and acts along the straight line which joins them. And he also shows that gravity acts on the bodies (with spherically symmetric mass distributions relative to their centers) in just the same way that it would act on point particles having the same masses as the spheres and located at their centers. This allowed the formulation of the n-body problem, which models a group of heavenly bodies: consider n point masses in three-dimensional Euclidean space, whose initial positions and velocities are specified at a time t_0, and suppose the force of attraction obeys Newton's Law of Universal Gravitation: how will the system evolve? This means we have to find a global solution of the initial value problem for the differential equation describing the n-body problem (Diacu and Holmes 1996: Chap. 1).

But here is the irony: the differential equations of the 2-body problem are easy to solve (Newton's difficulties with his own much more geometric formulation in Propositions XXXIX–XLI indicate the superiority of the idiom of differential equations here). However, for n larger than two, no other case has been solved completely. One might have thought that 'reducing' the models to differential equations would have made the solution of these centrally important problems about the solar system straightforward. But on the contrary, the equations articulated the complexity of the methods needed to solve problems in higher dimensional phase spaces, in order to express the physical situation (sub-systems of the solar system) accurately, as well as the great difficulty of finding complete solutions. For a 2-body problem, only one differential equation in one variable is needed, but for an n-body problem where $n > 2$, more than one differential equation, in more than one variable, is required. The severe difficulty of the n-body problem (not only in astronomy but in general mechanics) drove the development of physics for many decades. The work of Leibniz, Euler, Lagrange, Laplace and Hamilton replaced Newton's Laws with a single postulate, the Variational Principle, and replaced Newton's vectorial mechanics with an analysis in which the fundamental quantities are arbitrarily chosen coordinates and their time derivatives, and the dynamical relations are arrived at by a systematic process of differentiation. Lagrange's *Méchanique Analytique* (1788) introduced the Lagrangian form of the differential equations of motion for a system with n degrees of freedom, as we noted above, and generalized coordinates q_i ($i = 1, \ldots n$). That is, he showed that there is a form in which equations of motion can be cast, such that the form does not change no matter which variables one chooses. This innovation, the versatile use of variational principles, allowed physicists to choose whatever coordinates were most useful for describing the system, increasing the simplicity, elegance and scope of the mathematics (Fraser 2000: 93–101).

But of course in another obvious sense, the very complexity of the object, the solar system, forced the development of physics, since the solar system was the only thing that could be studied as a celestial mechanical system by the instruments available at the time. The main features of that complexity were already apparent to everyone: around the sun there are many planets, with moons around some planets. Uranus was identified by the important astronomer Herschel in 1781, and the asteroid belt between Mars and Jupiter was correctly identified at the beginning of the 19th century. Moreover, there were no important advances in telescopy until the mid-nineteenth century, so the controversies and advances apropos the mathematical models were notably theoretical/analytic. The culmination of these developments was the publication of Pierre-Simon Laplace's five-volume *Mécanique céleste* (1799–1825), where with immense mathematical skill he further elaborated these results into analytical methods for calculating the motions of the planets. In the early 1830s, in the context of general mechanics, William Rowan Hamilton discovered that if we regard a certain integral as a function of the initial and final coordinate values, this 'principal function' satisfies two first order partial differential equations; Carl Jacobi showed how to use Hamilton's approach to solve dynamical ordinary differential equations in terms of the Hamilton-Jacobi equation, later simplified and generalized by Alfred Clebsch (Diacu and Holmes: Chap. 4; see also Leech 1958: Chaps. 3–6).

Hermann von Helmholtz's publication of *On the Conservation of Force* (with *Kraft* defined in such a way that we would now translate it as energy) in 1847 was the culmination of efforts to find a mechanical equivalent for heat in the new domain of thermodynamics, and to aim at integrating a theory of mechanics, heat, light, electricity and magnetism by means of the notion of energy, rather than gravitational force (It was only in the 20th century that this aim became practicable, in the context of General Relativity Theory and Quantum Mechanics). Rudolf Clausius reformulated the work of Sadi Carnot and introduced the Second Law of Thermodynamics in 1850, as well as the notion of entropy in 1865. In an 1870 lecture entitled "On a Mechanical Theorem Applicable to Heat", he introduced the Virial Theorem, which states that in an assemblage of particles in gravitationally bound, stable statistical equilibrium, the average kinetic energy is equal to half the average potential energy (The import of the Virial Theorem within physics is more general, for it applies to any system in which the inter-particle forces are proportional to a gradient of a fixed power of the inter-particle distances). Whereas measuring the potential energy of a gravitational system requires the ability to measure its mass, measuring the kinetic energy depends on the measurement of the motions of bodies in the system. In the case of astronomical bodies, it was much easier to measure the latter than the former, so the Virial Theorem came to assume an important role in 20th c. cosmology, when it was applied to galaxies and galaxy clusters. However, in 1870, these objects were barely discernible: they were referred to as *nebulae*, clouds, because that was how they appeared. Many astronomers supposed that they would prove to have interesting internal structure, after Laplace in 1796, following the speculations of Kant, proposed the nebular hypothesis that the solar system emerged from a cloud of swirling dust. Thus the issue of models for the heavens reverted to the problem of reference, in the work of the astronomers Sir William Herschel and William Parsons, Earl of Rosse.

4 The Problem of Reference, from Rosse to Hubble

The path from the detection of 'nebulae' as cloudy smudges within the sole 'island galaxy' of the Milky Way to the recognition that many of them were in fact other galaxies far distant from our own, with complex internal structure encompassing hundreds of billions of stars is long and winding. Charles Messier catalogued the closest galaxy Andromeda as M31 in 1764, and William Herschel estimated that it was about 2000 times further away from us than Sirius (which is one of the stars closest to us). Herschel's large reflecting telescopes produced a dramatic increase in the ability of astronomers to watch the heavens; in 1789 he proposed that nebulae were made up of self-luminous nebulous material. He made hundreds of drawings of them, looking for significant morphological differences, or patterns of development, as he searched for evidence of his nebular hypothesis that clusters of stars may be formed from nebulae (Laplace modified the nebular hypothesis, as noted above, to speculate that the solar system was originally formed from a cloud of gases). Herschel's son John revised his father's catalogue for the Northern Hemisphere, and established a catalogue for the Southern Hemisphere as well, and kept alive the question of the composition of the nebulae: what were they made of? Alongside tabulations of positions, astronomical observations were drawn by hand; John Herschel was known for his meticulous sketches, which he hoped could be used in series, and by future astronomers, to determine change and motion in celestial configurations (Nasim 2010).

In 1845, William Parson, Earl of Rosse, built the largest telescope in the world: its speculum mirror was six feet in diameter, with a focal length of over four feet. He hoped to discover some of the fine structure of Herschel's nebulae. Soon after the telescope was set up, next to a smaller one that was equatorially mounted, he pointed it at Messier 51 (what we now call the Whirlpool Galaxy, a bright, face-on spiral with a companion) and discovered both its spiral, swirled structure and its companion. The discernment of the *shape* of the nebula was decisive. He sketched it repeatedly, in two steps: first he used the smaller telescope to scale the drawing, and then the large one to fill in the details. Herschel saw Rosse's sketches, presented at a meeting of the British Association for the Advancement of Science and was enthusiastically supportive. These drawings were later improved and engraved, so that the nebula was represented in negative, as black on a white background. So in Rosse's research project, the production of an astronomical image was an interplay between what was seen through the telescope, and what was carefully sketched by lamplight by Rosse and various assistants, thereafter to be re-fashioned as an engraving. This was the first (representational) model of a galaxy (Nasim 2010) (Fig. 3).

In the last two decades of the 19th century, astronomers solved various technical problems (how to keep the telescope and camera pointing in the right direction, over a period of time, for example) and profited from the introduction of dry plate photography, so that by 1887 a consortium of twenty observatories could produce a

Fig. 3 Rosse's Sketch of Messier 51

comprehensive astronomical atlas from photographic images. Comparison of photographs rapidly made clear how variable sketches had been as records of celestial objects, especially nebulae. Once astronomers had a firmer grasp of what they were trying to look at, the next step was to estimate how far away they were, and then to combine that knowledge with star counts and further estimations of stellar velocities within a given galaxy. Up to this point, the application of classical mechanics to these mysterious objects had really only been a pipedream. In the first decades of the 20th century it became a true research program, once nebulae were acknowledged to be extra-galactic objects, much larger and farther away than anyone in the 19th century suspected, and ever more powerful telescopes were able to track their motions and resolve their images. In the meantime, however, classical mechanics was being transformed, and the ensuing theoretical disputes affected the work of astronomers as well.

The development of Newtonian mechanics was therefore not ‘normal science’ in Kuhn’s sense. The emergence of electro-magnetic theory, the independent development of chemistry, and the study of thermodynamics were shaped by a growing awareness that in different domains forces other than gravity were important and demanded codification, and that the notation of differential equations, the study of symmetries, and the category of energy (as opposed to force) should be central to mechanics. However, the most direct challenge to Newtonian mechanics of course came from Einstein’s special and general theories of relativity, which explored the consequences of the equivalence of inertial frames (Special Relativity) and of accelerated frames (General Relativity), given the invariant speed of light. We have noted that Einstein proposed an equivalence between matter and energy, a 4-dimensional space-time continuum curved locally and perhaps globally by the matter and energy located in it, a dilation or slowing down of the passage of time registered by an observer in one reference frame as occurring in another reference frame moving (from the perspective of the first reference frame) close to the speed of light, and the notion of a light-cone as a formal limit on cosmic causal interaction. It was clear that these revisions of classical mechanics would have significant consequences for astronomy, certain aspects of which were beginning to change into modern scientific cosmology. In the late eighteenth and early 19th century, cosmology had remained merely speculative, driven by the metaphysical certainty of Leibniz and Goethe that in nature, ‘everything strives’. However, Relativity Theory did not impinge immediately on the study of galaxies. Rather, it was the characterization of ‘Cepheid variables’ by Henrietta Swan Leavitt and then the study of the ‘red shift’ of the electro-magnetic radiation emitted from stars by Edwin Hubble (both more closely related to problems of reference and taxonomy than to theoretical speculation) which moved the study of galaxies into the heart of modern cosmology.

The astronomer Edwin Hubble identified nebulae as galaxies, and then studied galaxies by analyzing the emission spectra of the light emitted by their stars; he noted that the standard patterns of spectral lines were for the most part shifted toward the red end of the spectrum (Some galaxies, like Andromeda which is in fact moving towards us, are blue-shifted). This he interpreted as a ‘Doppler shift’, which we know from ordinary experience as the lowering of the tone (due to the sound wave's perceived lengthening) of a train whistle when it rushes past us; that is, he took it as evidence that the galaxies (most of them) were receding from us. His famous law proposed in 1929 posits a linear relation defined by the Hubble constant between recessional velocity and distance, so that a measurement of red shift could be used to give an accurate estimate of how far away from us a galaxy lies. He also used 'standard stars' called Cepheid variables, whose period of variation and absolute luminosity are tightly related, as signposts; in combination, these factors allowed him to see that nebulae were extra-galactic, and to estimate their distances from us. Thus it was only during the 1920s that the scale of the universe began to dawn on astronomers (Liddle and Loveday 2008). In 1936, Hubble wrote in his influential book *The Realm of Nebulae* that "valuable information has been assembled concerning the scale of nebular distances, the general features of nebulae, such as their dimensions, luminosities, and masses, their structure and stellar contents, their large-scale distribution in space, and the curious velocity-distance relation" (Hubble 1982/2013: 181) (Fig. 4).

From that point on, scientists were puzzled about how to address the mismatch between astrophysical theory, originally based on the behavior of objects in the solar system, and the measurement of celestial systems. If rotating galaxies behaved like our solar system (which has a dominant central mass, the sun), the speed of stars at the outskirts of a galaxy should diminish in proportion to the reciprocal of the square root of their distance from the center; but stars on the outskirts of galaxies go much too fast. The laws of Newtonian physics predict that such high speeds would pull the galaxy apart. Thus in order to explain the stability of a

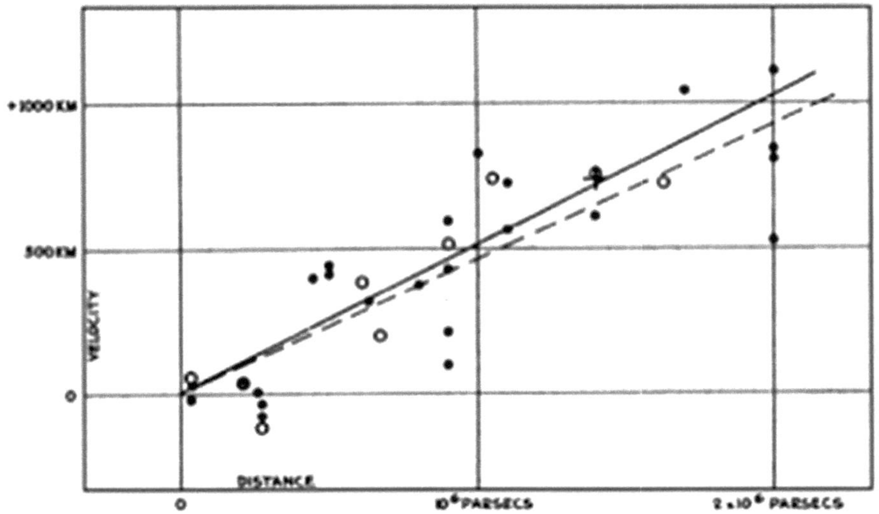

Fig. 4 Hubble's Date (1929)

galaxy, scientists either had to assume there is much more matter in a galaxy than we can see in the form of stars like our sun, or that Newton's laws must be revised for large systems.

5 Edwin Hubble and Fritz Zwicky

In 1937, the astronomer Fritz Zwicky took issue with Hubble on a number of points. He announced at the beginning of his paper "On the Masses of Nebulae and Clusters of Nebulae", that the determination of the masses of extragalactic nebulae was a central problem for astrophysics. "Masses of nebulae until recently were estimated either from the luminosities of nebulae or from their internal rotations", he noted, and then asserted that both these methods of reckoning nebular masses were unreliable. The adding up of observed luminosities gave figures that are clearly too low; and the models used for reckoning mass on the basis of observed internal motions were too indeterminate. Better models were needed, not least because Zwicky was convinced that in addition to luminous matter, galaxies (and the larger formations of galaxy clusters) included 'dark matter'. He wrote, "We must know how much dark matter is incorporated in nebulae in the forms of cool and cold stars, macroscopic and microscopic solid bodies, and gases" (Zwicky 1937). It would be anachronistic to read Zwicky here as supporting or even introducing the current hypothesis of 'dark matter', since he used the term simply to indicate that he thought that our telescopes cannot see some or most of what is actually included in a galaxy or galaxy cluster. There was luminous matter, which we can detect, and dark matter which (as yet) we can't. This made it all the more important to be able to estimate the mass of a galaxy or galaxy cluster on the basis of the internal movements of its visible components; thus we would have to improve upon the mechanical models used, so that those estimates could become more accurate. He discussed four kinds of models, the first of which, Hubble's model, he dismissed.

In *The Realm of Nebulae*, Hubble argued that from observations of internal rotations, good values of the mass of a galaxy should be derived. He wrote, "Apart from uncertainties in the dynamical picture, the orbital motion of a point in the equatorial plane of a nebula should be determined by the mass of material inside the orbit. That mass can be calculated in much the same way in which the mass of the sun is found from the orbital motion of the earth (or of the other planets)" (Hubble 1982/2013: 179). However, he expressed some doubts about how to interpret available data about both galaxies and galaxy clusters. Zwicky diagnosed the problem in terms of the indeterminacy of the mechanical model, for one could make the assumption that the 'internal viscosity' of a nebula was negligible, or that it was very great. In the former case, the observed angular velocities will not allow the computation of the mass of the system; in the latter case, the nebula will rotate like a solid body, regardless of what its total mass and distribution of that mass may be. For intermediate and more realistic cases, Zwicky argued, "It is not possible to derive the masses of nebulae from observed rotations without the use of additional information". If, for example, there were a central, highly viscous core with distant outlying, little-interacting components, one would need information about that viscosity and about the distribution of the outlying bodies. And he dismissed the analogy with the solar system as superficial.

Zwicky went on to propose three other possible models for calculating the mass of a galaxy or galaxy cluster. The second approach was to apply the Virial Theorem. If a galaxy cluster such as the Coma cluster was stationary, then "the virial theorem of classical mechanics gives the total mass of a cluster in terms of the average square of the velocities of the individual nebulae which constitute this cluster" (Zwicky 1937: 227). He argued that the Virial Theorem would work for the system, even if the nebulae are not evenly distributed throughout the cluster. But what if the cluster was not stationary? A brief calculation showed that, given the velocities, the Virial Theorem predicts that ultimately it will fly apart, which is odd, since then there should be no galaxy clusters at all; so there must be 'internebular material', whose nature and density should be further studied. Zwicky concluded that "the Virial Theorem as applied to clusters of nebulae provides for a test of the validity of the inverse square law of gravitational forces", because the distances are so enormous and these clusters are the largest known aggregates of matter (Zwicky 1937: 234). He also remarked that it would be desirable to apply the Virial Theorem to individual galaxies, but that it was just too difficult to measure the velocities of individual stars, as it was at that point in time. He treated this practical limitation as if he could not foresee its resolution.

The next model was that of gravitational lensing, a direct application of Einstein's theory of General Relativity; however, this was initially a merely speculative proposal, and wasn't tested and confirmed observationally until 1979. The final model was an extrapolation of ordinary statistical mechanics, "analogous to those which result in Boltzmann's principle". Zwicky's motivation in this section seemed to be to find a theory that would explain large-scale features of the universe without resorting to the kind of cosmological account (like the Big Bang theory, with which Hubble's Law became associated) he opposed, given his general disapproval of Hubble. Zwicky concluded, "It is not necessary as yet to call on evolutionary processes to explain why the representation of nebular types in clusters differs from that in the general field. Here, as in the interpretation of other astronomical phenomena, the idea of evolution may have been called upon prematurely. It cannot be overemphasized in this connection that systematic and irreversible evolutionary changes in the domain of astronomy have thus far in no case been definitely established" (Zwicky 1937: 239). For Zwicky, part of what was at stake was whether our model of the whole cosmos should be evolutionary or not.

Thus at the end of the 1930s, two important astronomers who had access to the same observational data on the largest material objects in the universe found themselves associated with two radically opposed views on the direction cosmology should take. Yet they were both equally puzzled by the discrepancy in estimates of the mass of these large objects. The evidence provided by star-counting or galaxy-counting, and the results of mechanically plausible models that calculate mass on the basis of the motions of stars in galaxies and of galaxies within clusters, simply did not agree. So the choice of theory could not be determined by observational results, and the clash of observational results could not be reconciled by theory. A quarter century later, astronomers were finally in a position to measure the velocities of components of a galaxy, and so to calculate the mass of the galaxy. Astronomers already had reliable evidence that a galaxy rotates about its center, based on the gradient in the stellar absorption lines on the major axis and the lack of such a gradient on the minor axis. If a galaxy were a mechanical system like the solar system, then we should expect that the velocity of its outer regions should decrease, as Kepler proposed and then, generalizing, Newton and Clausius demonstrated. The longer periods of revolution of Jupiter and Neptune, and the shorter periods of Mercury and Venus, can be accurately predicted. Even such a distinguished astronomer as Vesto Slipher (1875–1969) continued to characterize the radial velocity data of Andromeda and the Sombrero galaxy as "planetary" into the 1950s.

6 Vera Rubin and the Enigma of Andromeda

When Vera Rubin wrote her master's thesis in 1951, she reported her findings at a meeting of the American Astronomical Society, and met with little interest and some hostility. In the early 1960s, she and her graduate students made careful studies of the velocities of stars on the outskirts of Andromeda, because Rubin was interested in where galaxies actually end; they found that the galaxy rotation curve did not diminish, as expected, but remained flat. In her 1962 paper, she concluded "For $R > 8.5$ kpc, the stellar curve is flat, and does not decrease as is expected for Keplerian orbits". In 1970, she and W. Kent Ford, Jr. reported new data on Andromeda, profiting from the identification of almost 700 individual emission regions, as well as the use of image intensifiers that reduced observation times by a factor of 10. The edges of Andromeda did not move slower; they moved just as quickly as the inner regions (The Galaxy Andromeda is M31 in the Messier Catalogue) (Rubin and Ford 1970) (Fig. 4).

In 1980, with W. Kent Ford and Norbert Thonnard, she reported similar data for 21 further galaxies. While in the earlier papers she was reticent about drawing explicit conclusions, in this paper she writes, "Most galaxies exhibit rising rotational velocities at the last measured velocity; only for the very largest galaxies are the rotation curves flat. Thus the smallest Sc's (i.e. lowest luminosity) exhibit the same lack of Keplerian velocity decrease at large R as do the high-luminosity spirals. This form for the rotation curves implies that the mass is not centrally condensed, but that significant mass is located at large R. The integral mass is increasing at least as fast as R. The mass is not converging to a limiting mass at the edge of the optical image. The conclusion is inescapable that non-luminous matter exists beyond the optical galaxy" (Rubin et al. 1980). Since then, her observations have proved consistent with the measurement of velocities in a wide variety of other galaxies. Rubin has become more willing to say publicly that she thinks that the study of the kinematics of galaxies can teach us about galaxy evolution and cosmology; that galaxy dark halos must exist, and contain an order of magnitude more mass than the visible galaxy; and that galaxies remain an enigma. None of the models we have, in combination with received theories, can explain what we see as we watch the largest objects in the universe (Fig. 5).

We need to posit a spherical halo of dark matter around the spiral or ellipsoid that we see. This disparity also seems to hold for galaxy clusters. Thus its advocates claim that the evidence for dark matter is overwhelming; we are compelled to infer the existence of matter we can't see from its gravitational effects, and so too to search for novel ways to detect it, so that we can give it some characterization more determinate than simply "matter we can't see", Zwicky's tag for it 75 years ago. It is odd that in the context of this research program, the referential practice has so far been given the task of delineating more and more precisely a referential absence. Astronomers try to estimate precisely what percentage of matter is missing from this or that galaxy or galaxy cluster. No positive experimental program of detection has been successful: many candidates for dark matter have been entertained and ruled out, most recently neutrinos. Whatever it is, it bears a striking resemblance to 17th century 'aether', a perfect fluid that offers no resistance to anything passing through it, including and especially light, and no viscosity: it doesn't clump.

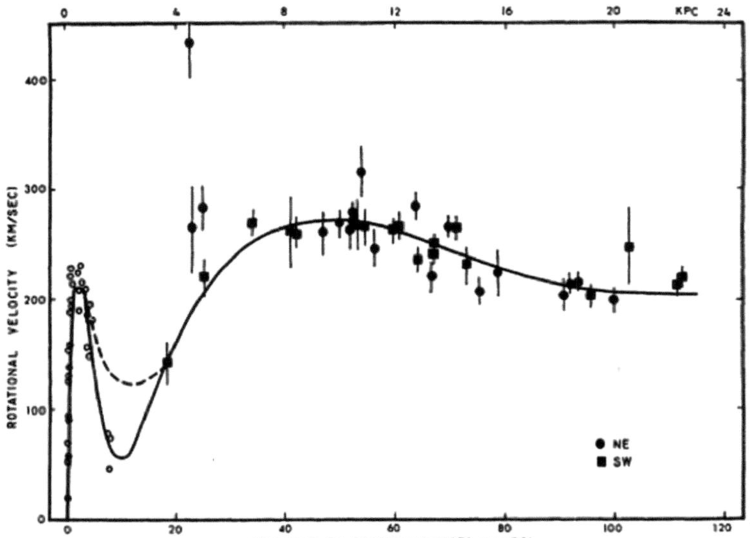

Rotational velocities for OB associations in M31, as a function of distance from the center. *Solid curve*, adopted rotation curve based on the velocities shown in Fig. 4. For $R \lesssim 12'$, curve is fifth-order polynomial; for $R > 12'$, curve is fourth-order polynomial required to remain approximately flat near $R = 120'$. Dashed curve near $R = 10'$ is a second rotation curve with higher inner minimum.

Fig. 5 Rubin's Data (1970)

Other scientists are unhappy with a scientific theory based on something that (up till now) resists detection altogether; they hold on to the hope of finding a strategy of unification, driven by observation rather than theory. The research program MOND (Modified Newtonian Dynamics) proposes that we revise Newtonian Mechanics to explain the uniform velocity of rotation of galaxies: perhaps Newton's Second Law of Motion ($F = ma$) is not universal, but applies only when gravitational acceleration is comparatively large. Perhaps for very low accelerations (like those experienced by stars on the outskirts of a galaxy) acceleration is not linearly proportional to force. Since its inception thirty years ago, proponents of MOND have tried various adjustments and refinements, trying to make it conform to General Relativity. However, other scientists are skeptical of the program because it looks like elaborate ad hoc reasoning, adding modifications to the theory simply to force it to accord with novel, more accurate observations. Proponents of MOND respond by observing that there is no reason why considerations of scale, in the study of such enormous objects, shouldn't prove pertinent to and so require expression in physical theory (see Sanders 2010).

7 Coda

I end the chapter with a final reflection, whose debt to Leibniz is acknowledged throughout this book. In general, science is (to state the obvious) both mathematical and empirical. In order to do good science, we must be able to refer successfully, so that we can show publicly and clearly what we are talking about. And we must analyze well—here I invoke Leibniz's definition of analysis as the search for conditions of intelligibility—to discover productive and explanatory conditions of intelligibility for the things we are thinking about. In order to evaluate whether our means of analysis are really productive and explanatory, we need to be able to denote—publicly and clearly—what we are considering. And in order to check whether our ways of referring are really public and clear, we must set the object of investigation in a more abstract discursive context where we can study it deeply and broadly. Sometimes one task is more difficult, sometimes the other, sometimes both. The tasks themselves are very different, so it is not surprising that they generate different kinds of discourse.

In referential discourse, we do our best to honor the extra-discursive world as what it is, with the best empirical means at our disposal; in analytic discourse, we treat the world in a sense as discourse and totalize, infinitize, simplify, or abstract it in the many ways that discourse allows. The advantage of analytic discourse is that it is great for organization, indexing and generalization; however, it also tends to unfocus the specificity of things, to make them ghostly. The advantage of referential discourse is that it does better justice to the rich, specific variety of things, but it often loses its way in the forest of research because of all those trees. In sum, science does its work best when we refer and analyze in tandem. However, the kinds of representations that make successful reference possible and those that make successful analysis possible are *not the same*, so that significant scientific and mathematical work typically proceeds by means of heterogeneous discourses that must be rationally reconciled without one collapsing into the other. The growth of scientific knowledge often stems from the work of reconciliation, whose fine structure has not received the attention it deserves. Philosophers need to pay more attention to the various ways in which scientists bring disparate discourses into rational, and productive, relation.

References

Cassirer, E. (1923/1953). *Substance and function and Einstein's theory of relativity* (W. C. Swabey & M. C. Swabey, Trans.). Chicago: Open Court.

Chemla, K., & Shuchun, G. (Trans.). (2004). *Les neuf chapitres: le classique mathématique de la Chine ancienne et ses commentaires*. Paris: Dunod.

Cohen, I. B. (1971). *Introduction to Newton's* Principia. Cambridge, MA: Harvard University Press.

Cook, A. H. (1998). *Edmond Halley: Charting the Heavens and the Seas*. Oxford: Oxford University Press.

Diacu, F., & Holmes, P. (1996). *Celestial encounters: the origins of chaos and stability*. Princeton: Princeton University Press.

Fraser, C. (2000). Hamilton-Jacobi methods and Weierstrassian field theory in the calculus of variations. In E. Grosholz & H. Breger (Eds.), *The growth of mathematical knowledge* (pp. 93–101). Dordrecht: Kluwer.

Grosholz, E. (1987). Some uses of proportion in Newton's *Principia*, Book I. *Studies in History and Philosophy of Science, 18*(2), 209–220.

Hendry, R. (2001). Mathematics, representation, and molecular structure. In U. Klein (Ed.), *Tools and modes of representation in the laboratory sciences* (pp. 221–236). Dordrecht: Kluwer.

Hubble, E. (1982/2013). *The realm of nebulae*. New Haven: Yale University Press.

Leech, J. W. (1958). *Classical mechanics*. London: Methuen.

Liddle, A., & Loveday, J. (2008). *Oxford companion to cosmology*. Oxford: Oxford University Press.

Nasim, O. W. (2010). Observation, working images and procedures: The 'Great Spiral' in Lord Rosse's astronomical records and beyond. *British Journal for History of Science, 43*(3), 353–389.

Newton, I. (1999). *The* Principia: *Mathematical principles of natural philosophy* (I. B. Cohen, A. Whitman, & J. Budenz, Trans.). Berkeley: University of California Press.

Rubin, V., & Ford, W. K. (1970). Rotation of the andromeda nebula from a spectroscopic survey of emission regions. *The Astrophysical Journal, 159*, 379–403.

Rubin, V., Ford, W. K., & Thonnard, N. (1980). Rotational properties of 21 Sc galaxies with a large range of luminosities and radii, from NGC 4605 (R = 4 kpc) to UCG 2885 (R = 122 kpc). *The Astrophysical Journal, 238*, 471–487.

Sanders, R. H. (2010). *The dark matter problem: An historical perspective*. Cambridge: Cambridge University Press.

Sylla, E. (1984). Compounding ratios. In E. Mendelsohn (Ed.), *Transformation and tradition in the sciences* (pp. 11–43). Cambridge: Cambridge University Press.

Voelkel, J. (2001). *Johannes Kepler and the new astronomy*. Oxford: Oxford University Press.
Wilson, R. N. (2014). *Reflecting telescope optics I: Basic design theory and its historical development*. New York: Springer.
Zwicky, F. (1937). On the masses of nebulae and of clusters of nebulae. *The Astrophysical Journal, 86*(3), 217–246.

Afterword

As a child, I learned about thrilling developments in 20th c. mathematics from reading *Scientific American* and science fiction (especially the works of Isaac Asimov and Ray Bradbury). I wanted to investigate the hypercube, and rethink the relation between space and time in terms of General Relativity Theory, and figure out how to turn a sphere inside out. In 1969 I met a few mathematicians who were doing research in Cambridge, Massachusetts, and listened in on their discussions at *The Blue Parrot*. I met others at the University of Chicago and Yale while I was earning my degrees, and still others in the early 1980s in Paris. In college and graduate school, since I couldn't figure out how to storm the ramparts of mathematics, I climbed over the walls of philosophy of mathematics, furnished with wysteria vines and the odd ladder. I dutifully took courses in formal logic (as I taught introductory logic and advanced logic at Penn State later on), and read the work of the philosophers collected in *Philosophy of Mathematics: Selected Readings* edited by Paul Benacerraf and Hilary Putnam (1964), *The Philosophy of Mathematics* edited by W. D. Hart (1996), and *The Philosophy of Mathematics: An Anthology* edited by Dale Jacquette (2002). The cast of characters hasn't changed much in half a century; yet I discuss very little of their work in the present book. Why is that?

The short answer is that what has always interested me in mathematics is invention, the way mathematics helps us to discover realms that we would never have discovered without its various items and languages, and its methods of abstraction and concretion, and its ways of formulating problems. (During my years at Penn State, I have come to regard physics, astrophysics and cosmology in the same way.) Early on, it was only in the writings of Imre Lakatos and Tobias Dantzig that I found some glimmer of the excitement I discerned, as through a glass darkly, in mathematical research: the discovery of the unexpected at the frontiers of knowledge. For me, that was precisely where all the philosophical interest was located. What struck me about those two books was that the authors not only made arguments, but told stories, historical narratives. History recounts the slow processes by which cultural traditions are formed, and then notes how the unexpected emerges from them, searching in retrospect to explain the disruptions and explore their consequences. Thus when I discovered the work of the friends and colleagues

E.R. Grosholz, *Starry Reckoning: Reference and Analysis in Mathematics and Cosmology*, Studies in Applied Philosophy, Epistemology and Rational Ethics 30, DOI 10.1007/978-3-319-46690-3

I discuss in the *Preface* to this book, starting with Donald Gillies (a student of Lakatos!) and Carlo Cellucci (a friend of Donald!) just about the time I started to teach at Penn State forty years ago, as well as the French tradition of philosophy of mathematics which had always included history, I saw the possibility of deeper exchange with other philosophers of mathematics.

And that exchange encouraged me to continue in two directions. The first was to go on studying mathematics and physics formally and informally. I'm grateful to my colleagues who let me sit in on their courses and who participated in the discussion groups and workshops I organized, or just spent some of their precious time talking to me and explaining things: Gordon Fleming, Winnie Li, Abhay Ashtekar, John Roe, Lee Smolin, Mihran Papikian, and (at Cornell) Roald Hoffmann. Given the highly specialized and peculiar nature of the languages of mathematics and science, a philosopher needs this kind of conversation and guidance: accuracy is very important. The second was to investigate the nature of time itself. "Time like an ever-rolling stream / Bears all its sons away / They fly, forgotten, as a dream / Dies at the opening day." So Isaac Watts once wrote of the asymmetry of time; but time also constantly brings forth the new and unexpected. The asymmetry of time is (to borrow Bergson's word) creative as well as fatal. Thus I regard the philosophical error of logicians, who wish to explain discovery away as an accident of human psychology, as akin to the philosophical error of those who think that the arrow of time can similarly be explained away. Both errors stem, it seems to me, from an impoverished view of rationality and from a fear of the unforeseeable. But this intellectual caution is unworthy of the philosophical tradition that gave us Descartes, Spinoza, Newton, Locke and Leibniz.

Appendix A
Historical Background for Fermat's Last Theorem

Analytic number theory uses real and complex analysis to study problems about the integers. In the 18th century Euler used analysis to re-prove the infinity of the prime numbers, and his approach then inspired Dirichlet, Jacobi and Riemann in the 19th century. This path of research led to important results about the distribution of primes in the work of Hadamard and de la Vallée-Poussin around 1900, because the reduction of questions about integers to questions about the divergence and convergence of certain series, offered much more powerful and flexible techniques than algebra in many cases. Conversely, once this habit of transposing problems upstairs to real and complex analysis was established, problems that arose originally in the infinitesimal calculus turned out to have important consequences for the study of the integers: the study of elliptic integrals began at the end of the 17th century in connection with the mathematical modeling of the pendulum, which entailed finding a way to determine the arc length of an ellipse. 18th c. mathematicians had the tendency to study problems of number theory analytically, embedding the study of the integers in the study of real-valued functions, and 19th c. mathematicians often embedded real analysis in complex analysis; this provides an important background for understanding the reduction of Fermat's Last Theorem to the Taniyama-Shimura Conjecture, the converse of the theorem of Eichler and Shimura.

I. From Euler to Riemann

Leonhard Euler studied the harmonic series in the mid-eighteenth century as a real-valued function, in order to prove in a quite non-Euclidean way that there are infinitely many primes. Knowing that the harmonic series (1 + 1/2 + 1/3 + 1/4 + ⋯ + 1/*n* + … for all natural numbers *n*) diverges, that is, its sum becomes arbitrarily large as more and more terms are added, he used this fact to argue that if there were only finitely many primes, then the unique factorization of positive integers into prime powers would entail that

$$\sum_{n=1}^{\infty} 1/n = \prod_p (1 + 1/p + 1/p^2 + \ldots) = \prod_p (1 - 1/p)^{-1}$$

E.R. Grosholz, *Starry Reckoning: Reference and Analysis in Mathematics and Cosmology*, Studies in Applied Philosophy, Epistemology and Rational Ethics 30, DOI 10.1007/978-3-319-46690-3

and thus a divergent series and a finite quantity would be equal, which proves the result by reductio ad absurdum. This line of thought led him to wonder whether the 'prime harmonic series'

$$(1+1/2+1/3+1/5+\cdots+1/p\ldots \text{for all primes } p)$$

also has an infinite sum. To answer the question by means of a problem reduction, he looked at the related sum

$$\zeta(s)=1+1/2^s+1/3^s+1/4^s+\cdots+1/n^s\ldots \text{for all integers } n.$$

This is the celebrated zeta function (given its name a century later by Riemann). He knew that if s is bigger than 1, the sum has a finite answer. Dividing this sum up into one part involving all the prime terms, and another part involving all the non-prime terms, he asked whether as s becomes closer and closer to 1 the first part, the sum

$$1+1/2^s+1/3^s+1/5^s+1/7^s\ldots,$$

increases without bound. This would mean that the sum

$$1+1/2+1/3+1/5+\cdots+1/p\ldots$$

for all primes p is in fact infinite. Once again, because of the unique prime decomposition of the natural numbers, this train of reasoning led him not only to the conclusion that $\zeta(s)$ does become infinite as s tends to 1, but in the process led him to formulate the equation,

$$\zeta(s)=1/(1-(1/2^s))\cdot 1/(1-(1/3^s))\cdot 1/(1-(1/5^s))\cdot\cdots\cdot 1/(1-1/p^s)) \\ \cdot\ldots \text{for all primes } p,$$

which of course links an infinite sum involving the natural numbers with an infinite product involving the primes p. There is nothing a number theorist likes better than bringing an additive decomposition into productive relation with a multiplicative decomposition!

These problems led number theorists to wonder if there is any way to characterize or quantify the distribution of the prime numbers within the integers: they seem to thin out as the integers get larger. Both Gauss (around 1790) and Dirichlet (around 1840) conjectured that the 'prime number function' $\pi(x)$ that counts the primes less than x (x is real), is asymptotically equivalent to the function $x/\ln x$. Dirichlet's investigations into the problem of the distribution of primes engendered a host of auxiliary notions, whose importance was borne out in the following 150 years. He introduced the character function, an arithmetic function that maps the integers to a field, here the real numbers. Intended to sort out the primes depending on the remainder they leave when divided by k, the function χ must satisfy these conditions:

(1) $\chi(n) = \chi(n + k)$ for all n (it is periodic with period k so that if $a \equiv b \pmod{k}$ then $\chi(a) = \chi(b)$); (2) $\chi(mn) = \chi(m)\chi(n)$ for all integers m, n (it is completely multiplicative); and (3) $\chi(n) = 0$ if n and k have a common factor and is non-zero otherwise, so $\chi(1) = 1$. Dirichlet used the character function to generalize the zeta series: his modified zeta series is known as the Dirichlet L-series, and has the form (where s is real and greater than 1),

$$L(\mathrm{s},\chi) = \chi(1)/1^{\mathrm{s}} + \chi(2)/2^{\mathrm{s}} + \chi(3)/3^{\mathrm{s}} + \chi(4)/4^{\mathrm{s}} + \cdots + \chi(n)/n^{\mathrm{s}} \ldots \text{for all } n.$$

Riemann studied the zeta function as a complex function of one variable w in his memoir "On the Number of Primes Less than a Given Magnitude" (1859), noting that for $\mathrm{Re}(w) > 1$, the series is absolutely convergent, and indeed converges uniformly in any region of the form $\mathrm{Re}(w) > -1 + \varepsilon$ for $\varepsilon > 0$. This means that it gives rise to an analytic function on the half plane $\mathcal{H}$, and that one can exploit the additive and multiplicative decomposition of the integers to write,

$$\zeta(s) = \sum_{n=1}^{\infty} n^{-s}$$
$$\zeta(s) = \Pi_{p'}(1 - \mathrm{p}^{-s})^{-1}$$
$$\log\zeta(s) = \Sigma_{p'} - \log(1 - p^{-s})$$
$$\log\zeta(s) = \Sigma_{p'} \sum_{n=1}^{\infty} p^{-ns}/n$$

(He also baptized the function with the Greek letter ζ.) All of these facts turned out to be very useful. Riemann proved that the zeta function is meromorphic on $\mathcal{H}$ with a simple pole at $s = 1$ of residue 1 and no other poles by using Abel's method of summation by parts, which is analogous to integration by parts.

Riemann also generalized the notion of character by allowing s and the numbers $\chi(n)$ to be complex numbers. In his memoir of 1859, Riemann treated the Dirichlet L-series given above as a function from the complex numbers to the complex numbers, $L(w, \chi)$, and established a number of important results. Euler's Theorem states that if a and n are coprime (without any common factors), $a^{\varphi(n)} \equiv 1 \pmod{n}$, where φ is the totient function which maps the natural numbers to the natural numbers: $\varphi(n)$ counts the number of primes less than n that are relatively prime to n. Riemann gave this theorem new meaning: since

$$\chi\left(a^{\varphi(k)}\right) = \chi(1) = 1,$$

and

$$\chi\left(a^{\varphi(k)}\right) = \chi(a)^{\varphi(k)},$$

then for all a relatively prime to k, $\chi(a)$ is a $\varphi(k)$th complex root of unity. Riemann showed as well that the L-function can be extended to a meromorphic function on the whole complex plane by analytic continuation, formulated its functional equation, and demonstrated a relation between its zeroes and the distribution of prime numbers. He also made several sweeping and unproved claims in that memoir, which inspired Hadamard to make them rigorous and more precise. In 1896, Hadamard and de la Vallée-Poussin exploited Riemann's treatment of the zeta function to prove the conjecture that the prime number function φ is asymptotically equivalent to the function $x/\ln x$. In the 20th century, the Dirichlet L-function was further generalized by the work of Emil Artin, Erich Hecke and Robert Langlands, and plays an important role in Wiles' proof.

Riemann moreover transformed the study of analytic (complex) functions by introducing the geometric and ultimately topological notion of a Riemann surface, what we would now call a one-dimensional complex manifold, locally homeomorphic to the complex plane. If f is an arbitrary analytic function (in general many-valued on the complex plane $\mathbf{C}$), there is a corresponding Riemann surface S on which it is single-valued. So the theory of analytic functions of a complex variable is coextensive with the theory of single-valued analytic functions on Riemann surfaces. Their study is in turn deepened, reorganized and generalized by the introduction of group theory in the late 19th century, applied to problems of real and complex analysis by Dedekind, Klein and Poincaré. In algebraic geometry an elliptic curve is defined as a 1-dimensional Abelian variety, a projective algebraic variety that is also always a group. In complex analysis applied to number theory, a modular form is defined in terms of a matrix group $SL_2(\mathbf{Z})$ that acts on the complex upper half plane $\mathcal{H}$. The theory of p-adic representations (which serve along with L-functions as important 'middle terms' in Wiles' proof) belongs to representation theory, which allows us to map almost anything with symmetries to linear transformations of vector spaces and then to groups of certain invertible matrices. Thus group theory too plays an essential role in the definition of the items to which Wiles refers in the formulation of his main result.

Here, I take a side-step to explain p-adic numbers and representation theory, developed by Kurt Hensel (1897) building on insights of Ernst Kummer. The importation of notions from topology has had important consequences, because topology allowed for the generalization of notions of distance and closeness. Thus we get from $\mathbf{Q}$ to $\mathbf{R}$ and $\mathbf{C}$ by a definition of closeness that stems from the real line, using the Euclidean norm. But we get from $\mathbf{Q}$ to the p-adic field $\mathbf{Q}_p$ by a definition of closeness that stems from congruence relations. Within the ring of integers $\mathbf{Z}$ there is a kind of fine structure precipitated by relations of congruence. There is the reorganization of the integers when they are sorted into subsets by modding out by a natural number n; the subsets form a group, and when n is a prime p they form a field. From this reorganization we can move to a sorting mod p^2, and then to a sorting mod p^3, and so on. These relations of congruence led to a novel conception of closeness, which in the 20th century came to play a key role in number theory and led to the definition of p-adic number fields: two integers are close to each other when they remain in the same congruence class not only mod p, but also mod p^2, mod p^3, and in general mod p^n. This sense of closeness can be generalized from the integers **Z** to the rationals **Q** by the notion of a p-adic valuation, so that we arrive at $\mathbf{Q}_p$. The move to $\mathbf{Q}_p$, arises only when mathematicians embed **Q** in **R** and then **R** in **C** when **C** is understood as the complex plane and complex analysis comes to be carried out on it, and then when **Q** is understood in relation to **Q**[i]. Once the topological notion of open set and metric space (and allied notions) are defined in their full generality, inspired by the study of holomorphic and meromorphic functions on the complex plane, and group theory has transformed our notion of congruences (at the same time as it transforms our notion of symmetries), one can embed **Q** in $\mathbf{Q}_p$.

Representation theory, which originally stems from the work of Frobenius (1896), studies a group by mapping it into a group of invertible matrices; it is a kind of linearization of the group. (Mathematicians are typically interested in the group of automorphisms of a system, often a highly infinitary system.) Another way to formulate this definition is to say that a representation of a group G, which maps elements homomorphically from G into a group of $n \times n$ invertible matrices with entries in (for example, though it could be other fields) **R**, sets up the action of G on the vector space $\mathbf{R}^\mathbf{n}$ and so directly yields an **R**G-module. Representation theory thus combines the results of group theory and linear algebra in a useful way. Symmetry is measured by the number of automorphisms under which an object or system remains invariant. The bigger the automorphism group, the more symmetrical the thing is. Every mathematical system has a symmetry group G, and certain vector spaces associated with the system turn out to be **R**G modules. (An **R**G module is a vector space over **R** in which multiplication of elements of the group G by elements of **R** is defined, satisfying certain conditions; there is a close connection between **R**G modules and representations of G over **R**).

II. Elliptic Curves and Modular Forms

The study of elliptic functions begins in one sense with the study of the pendulum, and the determination of the arc length of an ellipse; this leads to the study of elliptic integrals, which engaged Giulio Fagnano and Leonhard Euler in the early eighteenth century and Laplace and Legendre later on in that century. Elliptic integrals can be thought of as generalizations of the trigonometric functions; so for example the position of a pendulum could be computed by a trigonometric function as a function of time for small angle oscillations, but a full solution for arbitrarily large oscillations required the use of elliptic integrals. Elliptic functions (like those of Jacobi and Weierstrass) were discovered as inverse functions of elliptic integrals. Gauss studied elliptic functions of a complex variable in the early 19th century, but published few of his results. However, another tendril in this root system leads back to John Wallis, the champion of arithmetic. Having studied the 'geometric progression' a, ar, ar^2, … in which each term is obtained by multiplying its predecessor by a constant ratio, Wallis went on to study the 'hypergeometric progression,' in which the successive multipliers are unequal. (He used his influential method of interpolation to generate these multipliers: an example is the factorial sequence 1, 2, 6, 24….). Percolating through the work of Euler, Lagrange,

Stirling and Pfaff, it re-surfaced in Gauss' investigations (1812), and later Riemann's (1857), into the hypergeometric equation, which is characterized by the second-order ordinary linear differential equation,

$$x(1-x)\, d^2y/dx^2 + (c-(a+b+1)x)\, dy/dx - aby = 0.$$

This equation was used by Gauss, Abel, Jacobi and Kummer to study the transformation problem for elliptic functions.

In the work of Leibniz and the Bernoullis, the discovery of general methods for differentiation and integration led to the study of a whole new family of curves, transcendental curves, and their study by means of differential equations, which required a host of new methods for their solution, very different from (but fruitfully related to) the methods used to solve polynomial equations. These novel items and methods defined over the real numbers take on a new range of significance once they are defined over the complex numbers. Thus too, what began as the study of number series was transformed when it was embedded in problems of complex analysis, involving differential equations and functions defined on the complex plane; it generated new items that require new methods to classify and characterize them. That is, problems that arise about familiar items (the integers), when they are lifted up into complex analysis, engender new items that require serious taxonomical investigation and in many ways are just as 'concrete' as the integers.

One of the most important examples of this is the study of elliptic functions. In the latter part of the 19th century, Karl Weierstrass and Carl Jacobi characterized different, equally important kinds of elliptic functions, where f is defined as a doubly periodic, meromorphic function on the complex plane, satisfying the condition (where a and b are complex, non-zero numbers and a/b is not real),

$$f(z+a) = f(z+b) = f(z) \text{ for all } z \text{ in } \mathbf{C}.$$

The difference between the Jacobi and the Weierstrass elliptic functions is that Jacobi's has a double pole in the period parallelogram, and Weierstrass' has two single poles. The Weierstrass elliptic function, with periods ω_1 and ω_2, is defined as

$$\wp(z;\omega_1,\omega_2) = \frac{1}{z^2} + \sum_{(m,n)\neq(0,0)} \left\{ \frac{1}{(z+m\omega_1+n\omega_2)^2} - \frac{1}{(m\omega_1+n\omega_2)^2} \right\}.$$

This function and its first derivative are related by the formula

$$\wp'(z)^2 = 4\wp(z)^3 - g_2\wp(z) - g_3.$$

Here, g_2 and g_3 are constants; this relationship is given in terms of an elliptic curve over the complex numbers.

Modular forms and modular functions are analytic functions on the complex upper half plane $\mathcal{H}$. They are a generalization of the notion of an elliptic function; Klein's work, deploying the notion of group developed earlier by Evariste Galois and Camille Jordan, can be seen as an attempt to disentangle the theory of modular functions and modular transformations from the theory of elliptic functions, combining Dedekind's treatment of modular transformations—based on the idea of the lattice of periods of an elliptic function—with Galois group theory and the strategy of using Riemann surfaces, sorting out cases by the genus of the associated Riemann surface. The genus of the associated Riemann surface can be any positive integer n and, roughly, counts the holes in it. (A sphere, for example, has no holes, and a torus has one.) The definition of modular forms and modular functions is given in terms of a group action on the complex upper half plane $\mathcal{H}$. Looking backwards from Wiles' proof, we are interested mostly in a certain kind of modular form, and thus a certain kind of group action. In particular, $SL_2(Z)$, the Special Linear Group, is the set of all 2×2 matrices with determinant 1 and integer entries (with ordinary matrix multiplication and matrix inversion); it is called the modular group and acts on H. Modular forms are the functions that transform in a (nearly) invariant way under the action and satisfy a holomorphy condition. A modular function is a modular form without the condition that $f(z)$ is holomorphic at $i\infty$; instead, a modular function may be meromorphic at $i\infty$. A modular form of weight k for the modular group is holomorphic, and of course holomorphic at the cusp $i\infty$, and satisfies the equation, with c divisible by N,

$$f\left(\frac{az+b}{cz+d}\right) = (cz+d)^k f(z)$$

Every non-zero modular form has thus two associated integers, its weight k and its level N.

In Wiles' proof, one is concerned with modular forms of weight 2, which satisfy in addition an important condition that picks out the 'newforms' that arise at every level N; the character function χ here plays a significant role. Modular forms of any given weight and level constitute a vector space. Linear operators called Hecke operators act on these vector spaces; an eigenform is a modular form that is an eigenvector simultaneously for all the Hecke operators. Though modular form theory arises in complex analysis, its most important applications are in number theory; so there is an historical dialectic worthy of philosophical reflection that moves from arithmetic to complex analysis and back to arithmetic. We can understand Kronecker's constructivism or the project of a model theorist like Angus Macintyre as such a reflection.

Modular form theory is a special case of the theory of automorphic forms, developed by Henri Poincaré during the 1880s on the basis of earlier work by Lazarus Fuchs and in competition with Klein. An automorphic function is a function on a space, a function that is invariant under the action of some group. The theory of automorphic functions of a single variable is, historically, a union of analytic function theory and group theory. An analytic function f on the complex plane is automorphic with respect to a discontinuous group Γ of linear (fractional) transformations of the plane if *f* takes the same value at points that are equivalent under Γ. Thus we think of *f* as a function on the pertinent quotient space: since a quotient space is constructed by identifying certain points of a given space by means of an equivalence relation, and then treating them as one point of the new, quotient space, the action of a group on a space turns it into a new space whose points are the orbits of the original space induced by the group action. (The orbit O_x of a point *x* in a space X with respect to a group Γ is the set of points of X to which *x* can be moved by the action of Γ, and the set of all orbits of X under the action of Γ is called the quotient of the action or the orbit space or the coinvariant space.)

The most important domain for such an analytic (nonconstant) *f* is the complex upper half plane $\mathcal{H}$; then Γ is the discontinuous group of linear (fractional) transformations (or we may call them linear transformations). More concretely, if *f* satisfies the following functional equation, where V is an element of Γ (a real discrete group) and *z* is an element of $\mathcal{H}$, then it is an automorphic function on Γ: $f(Vz) = f(z)$ and so too $f(Vz) = (cz + d)^2 f'(z)$. Thus a modular form is a function defined in terms of a group Γ that acts on $\mathcal{H}$; this group maps the points of $\mathcal{H}$ to a modular curve whose points are representatives of the orbits of Γ (the points of $\mathcal{H}$ equivalent under the action of Γ), resulting in a quotient space that is a compact complex manifold very much like an algebraic variety, a Riemann surface of genus *n*.

In the mid-twentieth century, Robert Langlands shows how in general the Riemann-Roch theorem can be applied to the calculation of dimensions of automorphic forms; he also produces the general theory of Eisenstein series, demonstrating that all automorphic forms arise in terms of cusp forms and the residues of Eisenstein series induced from cusp forms on smaller subgroups. The condition of modularity is important because then the elliptic curve's *L*-function will have an analytic continuation on the whole complex plane, which makes Wiles' proof the first great result of the Langlands Program, and a harbinger of further results.

Appendix B
More Detailed Account of the Proof of Eichler and Shimura

Eichler and Shimura proved the following congruence between certain modular forms and elliptic curves. Let $f(z) = \Sigma c_n e^{2\pi inz}$ (summing over n greater than or equal to 1) be a cusp form, in particular a normalized newform of weight 2 for $\Gamma_0(N)$ such that $c_n \varepsilon \mathbf{Z}$. (Note that the holomorphic differential $f(z)$dz is invariant under the action of $\Gamma_0(N)$.) By $\Gamma_0(N)$ we mean a certain congruence subgroup of the modular group $SL_2(\mathbf{Z})$, which acts on the complex upper half plane by fractional linear transformations. Associated to $\Gamma_0(N)$ is the Riemann surface $Y_0(N)$ arising from the action of that congruence subgroup of $SL_2(\mathbf{Z})$ on the complex upper half plane, folding it over on itself; the points of $Y_0(N)$ are then the orbits of the subgroup $\Gamma_0(N)$. If we compactify $Y_0(N)$ at its cusps, the resulting compact Riemann surface is $X_0(N)$, and we can think of $f(z)$dz as living on that compact surface. Can we find an elliptic curve E_f defined over $\mathbf{Q}$, which corresponds to f?

We can do this by constructing a suitable Abelian subvariety A of $J_0(N)$, the Jacobian of $X_0(N)$, such that $J_0(N)/A = E_f$, with the help of Hecke operators. All the normalized newforms of weight 2 for $\Gamma_0(N)$ such that $c_n \varepsilon \mathbf{Z}$ can be thought of as a vector space on $X_0(N)$. Since such modular forms are not easy to express explicitly (as Eisenstein series, another kind of modular form, by contrast are), specifying a basis for the vector space cannot proceed by direct calculation. Instead, Hecke operators T_m can be defined on the vector space, and can be used to locate linearly independent eigenvectors that provide a basis for it. Supposing that m and N are mutually prime, each Hecke operator T_m belongs to the ring of endomorphisms of $J_0(N)$, End ($J_0(N)$). It's important to keep in mind that the Hecke operators are defined in terms of a vector space of differentials $f(z)dz$ which are isomorphic to the vector space of the $f(z)$ just defined, Riemann surfaces treated as topological spaces, cycles on those surfaces, integrals over those cycles, and associated homology groups. (I will not go into more detail here.)

The Hecke operators constitute a Hecke algebra T over $\mathbf{Q}$, generated by the identity operation and the T_m. The Hecke algebra acts on a newform $f = \Sigma c_n e^{2\pi inz}$ by $T_m f = c_m f$. Thus, given a newform f with normalized $c_1 = 1$ and $c_n \varepsilon \mathbf{Z}$, we can define a one-dimensional representation ρ_f: T $\rightarrow \mathbf{Q}$, where $\rho_f(T_m) = c_m$ and $\rho_f(id) = c_1 = 1$. Since T can be decomposed into a direct sum of fields k_i of finite degree over Q, plus the nilradical (T) (elements in the algebra that are nilpotent, that

E.R. Grosholz, *Starry Reckoning: Reference and Analysis in Mathematics and Cosmology*, Studies in Applied Philosophy, Epistemology and Rational Ethics 30, DOI 10.1007/978-3-319-46690-3

is, where for some $n \varepsilon \mathbf{Z}$, $x^n = 0$), we can specify further that ρ_f(nilradical (T)) = 0 and that $\rho_f \mid k_i$ is a homomorphism from k_i to $\mathbf{Q}$. Each T_m is represented by a $g \times g$ matrix with entries in $\mathbf{Z}$, so each T_m is defined over $\mathbf{Q}$, and T is a finite-dimensional $\mathbf{Q}$-algebra and it is commutative. Then we can finally define A as the $\mathbf{Q}$-rational Abelian subvariety of $J_0(N)$ generated by the images $\alpha(J_0(N))$ for all $\alpha \varepsilon$ (ker $\rho \cap$ End $(J_0(N))$). So there it is, our A, but what is it exactly? Since A is the key middle term in this whole argument, it has to be properly constructed and then shown to do the job it is supposed to do.

Call E_f the quotient Abelian variety $J_0(N)/A$ which is $\mathbf{Q}$-rational, then, and call v the natural map from $J_0(N)$ to E_f, which is $\mathbf{Q}$-rational. We identify $J_0(N)$ with the Cartesian product of g copies of $\mathbf{C}$ (thus a space of dimension $2g$ over the field of real numbers) modded out by a lattice Λ of rank $2g$. The tangent space at its distinguished element is a Lie algebra J that can be identified with $\mathbf{C}^g$; the natural map from J = $\mathbf{C}^g$ to $\mathbf{C}^g/\Lambda$ is the exponential map. (Recall that an elliptic curve is a smooth, projective algebraic curve of genus 1 with a distinguished point O, as well as an Abelian variety of dimension 1 where O serves as the identity.) The Lie algebra of the Abelian subvariety A, A, is the tangent space at the distinguished element of A; it turns out to be a subalgebra of J. If we then carefully define $\mu(f)$, a linear functional on the Lie algebra J which maps $\mathbf{C}^g$ back down to $\mathbf{C}$, in terms of v, we can show that the ker $\mu(f)$ is A and so $\mu(f)$ usefully maps Λ to Λ_f, and that the elliptic curve E_f is isomorphic to $\mathbf{C}/\Lambda_f$.

Having constructed $J_0(N)/A = E_f$, we want to be able to show that the L-functions of the modular form f we started out with, and the elliptic curve E_f we arrived at, agree: $L(E_f, s) = L(f,s)$. An L-function can be represented by a sum taken over the positive integers where each factor is c_n/n^s, with s a complex variable restricted to some right half plane so the series converges absolutely. An L-function can also be represented by an Eulerian factorization, a product taken over the primes p in which c_p appears in every factor, combined with other items. The upshot of this successful identification is that the system of eigenvalues c_p of the newform f (considered as an eigenform of the Hecke operators T_p) that occur on the right side of the above equation, can be identified with the solution counts of the equation (mod p) that gives the elliptic curve E_f. Recall that an elliptic curve E is defined by a certain cubic equation, $y^2 = 4x^3 - g_2x - g_3$, where g_2 and g_3 are integers and $g_2^3 - 27g_3^2 \neq 0$. For each prime number p, we can associate a number $c_p(E)$ that counts the number of integer solutions (x, y) of E mod p. These are the coefficients that occur on the left side of the above equation.

In sum, the Jacobian of a modular curve is analogous to a complex elliptic curve in that both are complex tori and so have Abelian group structure. The celebrated Taniyama-Shimura modularity theorem says that every elliptic curve defined over Q is a homomorphic image of some Jacobian $J_0(N)$ and omit the next sentence. Only weight 2 eigenforms with Hecke eigenvalues in $\mathbf{Z}$ correspond to elliptic curves defined over $\mathbf{Q}$; more general eigenforms correspond to Abelian varieties of higher dimension. All the ways of explaining this correspondence involve appeals to 'large' structures to which the rather modest newform $f(z)$ can be lifted, investigated, and then re-deposited with a new affiliation, to an elliptic curve.

These appeals are clearly ampliative. We first define it in terms of the group of fractional linear transformations of the complex upper half plane, and the projective special group of matrices $SL_2(\mathbf{Z})$. We map the complex upper half plane onto itself, adding points at infinity to make a compact Riemann surface $X_0(N)$, which is a topological space as well as a geometrical space. On the space of weight 2 cusp forms we define a set of Hecke operators, which form a commutative algebra of operators. The space of weight 2 cusp forms $f(z)$ for $\Gamma_0(N)$ can be identified with the space of differentials $f(z)dz$ on $X_0(N)$. We make use of the Jacobian of $X_0(N)$, $J_0(N)$. Inspired by $J_0(N)$, we invoke $\mathbf{C}^g$ and a lattice Λ, construct their quotient $\mathbf{C}^g/\Lambda$, identify it with $J_0(N)$, and then invoke a tangent space at the distinguished element which is a Lie algebra J.

The process of going back down, however, is also nontrivial and requires further amplification. The study of matrices depends on the mapping of the matrix to an informative single number, the determinant or the trace for example. We defined a one-dimensional representation ρ (representations map groups to groups of invertible matrices) that maps T down to $\mathbf{Q}$, and the map v that maps $J_0(N)$ to E_f, and $\mu(f)$ that maps $\mathbf{C}^g$ back down to $\mathbf{C}$. And in the present context, an L-function takes either a modular form or an elliptic curve and reduces it to a set of positive integers. Indeed, the expression of $f(z)$ as a Fourier series, $f(z) = \Sigma c_n e^{2\pi inz}$ (summing over n greater than or equal to 1), does the same; when f is an eigenform for the Hecke operators, these coefficients turn out to be the eigenvalues of the associated Hecke operators.

Glossary

Automorphism An automorphism is a mapping of an object to itself, which preserves all its structure, so it is a bijective homomorphism (if the object can be construed as a set). The set of all automorphisms of an object is a group, the automorphism group, which in a general sense expresses the symmetry of the object. Roughly, the larger the automorphism group, the more symmetry the object has.

Character A **multiplicative character** (or **linear character**, or simply **character**) on a group G is a group homomorphism from G to the multiplicative group of a field (Artin 1966), usually the field of complex numbers. If G is any group, then the set Ch(G) of these morphisms forms an Abelian group under pointwise multiplication. This group is referred to as the character group of G. Sometimes only *unitary* characters are considered (thus the image is in the unit circle); other such homomorphisms are then called *quasi-characters*. Dirichlet characters can be seen as a special case of this definition. Multiplicative characters are linearly independent.

Chinese Remainder Theorem Let r and s be relatively prime positive integers (that is, they have no common divisors except 1), and let a and b be any two integers. Then we can always find an integer n such that $n \equiv a \pmod{r}$ and also $n \equiv b \pmod{s}$. Moreover, n is uniquely determined modulo $r \cdot s$. This result can be generalized in terms of a set of simultaneous congruences $n \equiv a_i \pmod{r_i}$.

Class Group The extent to which unique factorization fails in the ring of integers of an algebraic number field (or more generally any Dedekind domain) can be described by a certain group known as an ideal class group (or class group). If this group is finite (as it is in the case of the ring of algebraic integers of an algebraic number field), then the order of the group is called the **class number**. The class group of $\mathbf{O}_K$ (the set of all algebraic integers in K) is defined to be the group of fractional $\mathbf{O}_K$-ideals modulo the subgroup of principal fractional $\mathbf{O}_K$-ideals, where K is an algebraic number field. When $\mathbf{O}_K$ is a principal ideal domain, or when we can define a Euclidean norm on $\mathbf{O}_K$, this class group is trivial, that is, it is just the group of the element 1. More generally, the multiplicative theory of a Dedekind domain (whose class group may be infinite) is

E.R. Grosholz, *Starry Reckoning: Reference and Analysis in Mathematics and Cosmology*, Studies in Applied Philosophy, Epistemology and Rational Ethics 30, DOI 10.1007/978-3-319-46690-3

intimately tied to the structure of its class group. For example, the class group of a Dedekind domain is trivial if and only if the ring is a unique factorization domain.

Conductor The **conductor** of a ring is an ideal of a ring that measures how far it is from being integrally closed. The **conductor** of an abelian variety defined over a local or global field measures how bad the bad reduction is at some prime p.

Cyclotomic Field A cyclotomic field is a field constructed by adjoining a complex number that is a primitive root of unity to the field of rational numbers **Q**. Thus the nth cyclotomic field $\mathbf{Q}(\zeta_n)$ is constructed by adjoining a primitive nth root of unity to **Q**: $\mathbf{Q}(i)$ is the 4th cyclotomic field, because $i = \zeta_4$, since $i \cdot i \cdot i \cdot i = (-1) \cdot (-1) = 1$. Since the roots of unity form a cyclic group, adjoining the primitive root of unity will yield all the roots. A cyclotomic field is always a Galois extension of the field **Q**.

An **elliptic function** f is a doubly periodic, meromorphic function defined on the complex plane, which satisfies the condition (where a and b are complex, nonzero numbers and a/b is not real), $f(z+a) = f(z+b) = f(z)$ for all z in **C**. There are two kinds of canonical elliptic functions, the Jacobi elliptic function, which involves the theta function [Jacobi forms are related to double periodicity and further generalize modular forms], and the Weierstrass elliptic function. The centrally important elliptic functions are the Weierstrass $\wp$-function and its derivative $\wp'$. When we expand $\wp$ in a power series of its variable u, then the coefficients are modular forms; in fact, they are Eisenstein series for $SL_2(\mathbf{Z})$. This is the connection between elliptic functions and modular forms in general. Elliptic functions were discovered by Abel, Gauss and Jacobi in the 1820s.

An **elliptic curve** over a field K is a genus 1 curve defined over K with a point O in K. We can say both that an elliptic curve is a smooth, projective albraic curve of genus one with a distinguished point O, and that it is an abelian variety of dimension 1 where O serves as the identity. Take K to be the complex numbers: elliptic curves defined over **C** correspond to embeddings of the torus in the complex projective plane. If a modular curve has genus 1, then it is an elliptic curve over a suitable field K, chosen so that the curve contains a point with coordinates in K. When a modular curve $X(\Gamma)$ has genus 1, it is an elliptic curve over **C**: in this case the quotients of modular forms of the same weight give rise to meromorphic functions on $X(\Gamma)$, which are elliptic functions.

A **complex elliptic curve**. is a quotient of the complex plane by a lattice. It is an Abelian group, a compact Riemann surface, a torus, and a bijective correspondence with the set of ordered pairs of complex numbers satisfying a cubic equation of the form E.

$$\mathrm{E} : y^2 = 4x^3 - g_2 x - g_3$$

where g_2, g_3 are complex numbers and $g_2^3 - 27g_3^2$ does not equal zero.

The Modularity Theorem associates to E an **eigenform** $f = f_E$ in a vector space of weight 2 modular forms at a level N called the **conductor** of E. The eigenvalues of f are its **Fourier coefficients**. The Modularity Theorem only applies to elliptic curves defined over **Q**, which has a model given by $y^2 = x^3 - ax - b$ with a, b in **Q** and $4a^3 - 27b^2$ nonzero, which differs from the above model by the integer 4 in front of x^3.

Galois Group and Galois Extension A Galois extension E of a field F is an extension such that E/F is algebraic, and the field that is fixed by the group of automorphisms of E (that is, Aut(E/F)) is precisely F. Thus, for example, the Galois group of $\mathbf{Q}(\zeta_4) = \mathbf{Q}(i)$ consists of the automorphisms of $\mathbf{Q}(i)$ that permute the two elements i and $-i$ while leaving **Q** fixed.

Hecke Operator The modular forms of any given weight and level form a vector space; linear operators called **Hecke operators** (including the operator T_p for each prime p) act on these vector spaces.

Ideal An ideal C in a ring A is a non-empty subset of A with the properties that c_1 and c_2 in C imply that $c_1 - c_2$ is in C, and that c in C and a in A imply that ac and ca are in C. (Given any homomorphism of a ring A, the set of elements that map to zero is an ideal in A).

Fractional Ideal Let R be an integral domain and let K be its field of fractions, which is the smallest field in which it can be embedded: **Q** is the field of fractions of the ring/integral domain **Z**. Recall that a module over a ring is a generalization of a vector space over a field: if M is a left (resp. right) R-module and N is a subgroup of M, then N is a an R-submodule if, for any n in N and any r in R, the product rn (resp. nr) is in N. Then a fractional ideal of R is an R-submodule of K such that there exists a nonzero $r \in R$ such that $rI \subseteq R$. The element r can be thought of as clearing out the denominators in I.

Fundamental Domain If Γ is a real discrete group, the relation of Γ-equivalence partitions **H** into disjoint orbits Γz. A subset F of **H** that contains exactly one point from each orbit is called a fundamental set for Γ relative to **H**. (You can always find such a set, but it is not unique.) The most familiar groups have very simple fundamental sets. For example, the doubly periodic group is a parallelogram with two adjacent open sides and their common vertex adjoined. In each case, the fundamental set is an open set with some of its boundary points adjoined, and the complete boundary consists of line segments or circular arcs. Every discrete group admits a fundamental set of this kind.

A **group action** of a group G on a set X is a map $G \times X \to X$ where $(g, x) \to gx$ such that the identity element in G applied to any x in X goes to x, and $g_1g_2(x) = g_1(g_2x)$ for all g_1 and g_2 in G and all x in X. The **orbit** of a point x in X under the action of G is the set of points O_x s.t. $O_x = \{gx | g \in G\}$ and the **stabilizer** of a point x in X is the subgroup G_x s.t. $G_x = \{g \in G | gs = s\}$. Orbits and stabilizers are closely related. That is, the orbit of x in X under the action of G can be identified with the cosets of G_x in G, that is, the quotient G/G_x. This is just a set isomorphism.

Ideals: Prime Ideal An ideal P of a commutative ring R is prime if it has the following two properties: If a and b are elements of R such that their product ab is an element of P, then a is in P or b is in P, and P is not equal to the whole ring R. **Principal Ideal**. If b is an element of a commutative ring A with unity, the set (b) of all multiples xb of b, for any x in A, is a principal ideal. **Principal Fractional Ideal**. The principal fractional ideals are those R-submodules of K generated by a single nonzero element of K.

Ideal Class Group An ideal class group is the quotient group J_K/P_K where J_K is the set of all fractional ideals of K and P_K is the set of all principal ideals of K. The extent to which unique factorization fails in the ring of integers of an algebraic number field (or more generally any Dedekind domain) is registered by the ideal class group. If this group is finite (as it is in the case of the ring of integers of a number field), then the order of the group is called the class number.

Integral Domain A nonzero commutative ring in which the product of any two nonzero elements is nonzero; this is a generalization of the ring of integers, between the ring of integers and the field of the rationals.

Jacobian The Fréchet derivative in finite-dimensional spaces is the usual derivative. In particular, it is represented in coordinates by the **Jacobian matrix**, the matrix of all first-order partial derivatives of a vector valued function with respect to another vector. That is, we have $F: R^n \to R^m$, and this function is given by m real-valued component functions, $y_1(x_1, \ldots, x_n), \ldots, y_m(x_1, \ldots, x_n)$. The partial derivatives of all these functions (if they exist) can be organized in an m-by-n matrix. The **Jacobian** is the determinant of the Jacobian matrix if $m = n$.

Kronecker-Weber Theorem Every cyclotomic field is an abelian extension of **Q**, the field of rational numbers. This theorem provides a partial converse: every abelian extension of **Q** is contained in some cyclotomic field; thus, every algebraic integer whose Galois group is abelian can be expressed as a sum of roots of unity with rational coefficients.

L*-function; Dirichlet Characters** These are certain arithmetic functions which arise from completely multiplicative characters on the units of **Z**/*kZ**. Dirichlet characters are used to define Dirichlet L-functions, which are meromorphic functions with a variety of interesting analytic properties. If χ is a Dirichlet character, one defines its Dirichlet L-series by

$$L(s, \chi) = \sum_{n=1}^{\infty} \frac{\chi(n)}{n^s}$$

where s is a complex number with real part >1. By analytic continuation, this function can be extended to a meromorphic function on the whole complex plane. Dirichlet L-functions are generalizations of the Riemann zeta-function and appear prominently in the generalized Riemann hypothesis

Modular Forms These are complex analytic functions on the complex upper half plane **H**. A matrix group called the **modular group** acts on **H**, and modular forms are the functions that transform in a (nearly) invariant way under the action and satisfy a holomorphy condition. We can restrict the action to subgroups of the modular group called **congruence subgroups**, which gives rise to further modular forms. Modular form theory arises in complex analysis, but its most important applications are in number theory. Modular form theory is a special case of the theory of **automorphic forms**. A modular form of weight k for the modular group (isomorphic to $SL_2(Z)$) is holomorphic, and indeed holomorphic at the cusp i ∞, and satisfies the equation

$$f\left(\frac{az+b}{cz+d}\right) = (cz+d)^k f(z)$$

Every nonzero modular form has two associated integers, its weight k and its level N. The modular forms of any given weight and level form a vector space. Linear operators called **Hecke operators** (including the operator T_p for each prime p) act on these vector spaces. An **eigenform** is a modular form that is an eigenvector simultaneously for all the Hecke operators. A **cusp form** is a modular form which is holomorphic at all cusps and vanishes at all cusps.

The **modular group** $\Gamma$is a group of linear fractional transformations of the complex upper half plane, sending z to $az+b/cz+d$ where a, b, c and d are integers and $ad - bc = 1$. This is isomorphic to the special linear group $SL_2(\mathbf{Z})$, the set of all 2×2 matrices with integer entries with det = 1, modulo $\pm$ the identity matrix, the so-called $PSL_2(\mathbf{Z})$.

A **modular curve** is a quotient of the complex upper-half plane by the action of a congruence (or, more generally, a finite index) subgroup of $SL_2(\mathbf{Z})$. Every modular curve is a **Riemann surface**, as well as the corresponding algebraic curve, constructed as a quotient of the complex upper-half plane by the action of a congruence subgroup of the modular group $SL_2(\mathbf{Z})$. The term modular curve can also be used to refer to the compactified modular curve $X(\Gamma)$, obtained by adding finitely many points, the **cusps** of Γ, to this quotient, via an action on the extended complex upper-half plane. A modular curve can have any genus n. Points of a modular curve parametrize isomorphism classes of elliptic curves, together with some additional structure depending on the group Γ. This interpretation allows one to give a purely algebraic definition of modular curves, without reference to complex numbers, and moreover proves that modular curves are defined either over the field **Q** of rational numbers, or a cyclotomic field. The latter fact and its generalizations are of fundamental importance in number theory.

A **modular function** takes the complex upper-half plane to the complex numbers. Not all modular functions are modular forms. A modular function is a modular form without that condition that $f(z)$ is holomorphic at infinity (or cusps); instead, a modular function may be meromorphic at infinity, that is, it may have poles at cusps.

***P*-adic Numbers** The ***p*-adic number system** for any prime number p extends the ordinary arithmetic of the rational numbers in a way different from the extension of the rational number system to the real and complex number systems. The extension is achieved by an alternative interpretation of the concept of absolute value. The p-adic numbers were first described by Kurt Hensel in 1897, though with hindsight some of Kummer's earlier work can be interpreted as implicitly using p-adic numbers. The p-adic numbers were motivated primarily by an attempt to bring the ideas and techniques of power series methods into number theory. The field of p-adic analysis essentially provides an alternative form of calculus. For a given prime p, the field $\mathbf{Q}_p$ of p-adic numbers is a completion of the rational numbers. The field $\mathbf{Q}_p$ is also given a topology derived from a metric, which is itself derived from an alternative valuation on the rational numbers. This metric space is complete in the sense that every Cauchy sequence converges to a point in $\mathbf{Q}_p$. This is what allows the development of calculus on $\mathbf{Q}_p$, and it is the interaction of this analytic and algebraic structure which gives the p-adic number systems their power and utility. See also Appendix A.

Quadratic Reciprocity Theorem If p and q are distinct odd primes, the congruences $x^2 \equiv q(\bmod p)$ and also $x^2 \equiv p(\bmod q)$ are both solvable in $\mathbf{Z}$ unless both p and q leave the remainder 3 when divided by 4, in which case one of the congruences is solvable but the other is not. Here is another way to formulate this result. It involves the Legendre symbol $(^p/_q)$, used in the following way: $(^p/_q) = 1$ in case $x^2 \equiv p(\bmod q)$ is solvable for x, and $(^p/_q) = -1$in case $x^2 \equiv p(\bmod q)$ is not solvable for x. Then we write, $(^p/_q)(^q/_p) = (-1)^{(p-1)(q-1)/4}$.

Reciprocity Laws Let $f(x)$ be a monic irreducible polynomial with integral coefficients, and let p be a prime number. If we reduce the coefficients of $f(x)$ modulo p, we get a polynomial $f_p(x)$ with coefficients in the field $\mathbf{F}_p$ of p elements. The polynomial $f_p(x)$ may then factor, even though the original polynomial $f(x)$ was irreducible. If $f_p(x)$ factors over $\mathbf{F}_p$ into a product of distinct linear factors, we say that $f(x)$ splits completely mod p and we define **Spl(*f*)** to be the set of all primes such that $f(x)$ splits completely modulo p. Then the general reciprocity problem is, given $f(x)$ as above, describe the factorization of $f_p(x)$ as a function of the prime p. We also ask the related question: is there a rule that determines which primes belong to **Spl(*f*)**? It turns out that the reciprocity problem has been solved for polynomials that have an Abelian Galois group, but not for polynomials whose Galois group is not Abelian.

Recursive Set A set of natural numbers is called a recursive set if there exists an algorithm that terminates after a finite amount of time, and correctly determines whether or not a given number belongs to the set.

A **Riemann Surface** is a 1-dimensional complex manifold. A 2-dimensional real manifold can be turned into a Riemann surface only if it is orientable and metrizable; thus a sphere and a torus can be treated like a Riemann surface, but a Moebius strip, a Klein bottle and the real projective plane cannot

Representation Theory See Appendix A.

Zeta Function See Appendix A.

MIX
Papier aus verantwortungsvollen Quellen
Paper from responsible sources
FSC® C105338

If you have any concerns about our products,
you can contact us on
ProductSafety@springernature.com

In case Publisher is established outside the EU,
the EU authorized representative is:
Springer Nature Customer Service Center GmbH
Europaplatz 3, 69115 Heidelberg, Germany

Printed by Libri Plureos GmbH
in Hamburg, Germany